Advance Praise for *Do YOU Mean Business?*

"In 15 years of recruiting civil engineers I have found the most successful Business Line Leaders and CXOs are the ones who artfully blend engineering knowledge with business development skills; not an easy task. Babette Ten Haken's book, *Do YOU Mean Business?* challenges the stereotypes of engineers and business development professionals, who dwell under the same roof but operate in different silos. Babette provides a roadmap to achieve new levels of success by opening the lines of communication, comprehending customer needs, and working collaboratively in order to secure a higher percentage of wins and successful projects."

~MATT BARCUS, *Managing Principal, Precision Executive Search, and Managing Partner, CivilEngineeringCentral.com*

"Engineers who master the art of sales and marketing are a highly valuable asset. Imagine if 80% of your engineers had a good understanding of sales and marketing. Your company just might be that powerhouse you envision."

~ NANCY NARDIN, *Founder, Smart Selling Tools*™

"Engineers in the business world often struggle with the dilemma of getting a technical Master's/PhD or an MBA. Before making that decision, I encourage them to read *Do YOU Mean Business?* This book clearly outlines the roles of an engineer who is looking to grow their business acumen. The book also serves as an excellent blueprint for someone looking to break the status quo of their career. Finally, this book should be read by all individuals in an organization: it breaks down silos and allows teams to focus on the bottom line…providing exceptional value to customers. *Do YOU Mean Business?* is a 'must have' resource for all individuals (technical and non-technical) looking to grow their careers."

~ BEN MATTHEWS, P.E., *Project Manager, Global Engineering and Design Consultancy industry*

"Babette takes the seemingly complex and makes it easy to understand, and more important, easy to implement as it relates to creating and growing a business. In *Do YOU Mean Business?*, you'll find the perfect marriage between technical and non-technical, between theoretical and practical, as it relates to crafting a plan, building a team, and bringing business ideas to life. Read this book, implement her strategies, and you will learn to focus on what really matters."

~ SAM RICHTER, *best-selling author and speaker, author of* Take the Cold Out of Cold Calling *and the* Know More! *Center, Toolbar, and Newsletter*

"In my 12 years in the evolving healthcare information technology industry I had yet to find a book clearly outlining the benefits and approach of cross-functional communication and business to business strategy. Until now. *Do YOU Mean Business* should be the required 'field manual' for any entrepreneur, sales or engineering leader. Now I know how to make engineers my best advocates and partners in creating value for my clients. Stop leaving mixed messages and money on the table. Get moving with this plan!"

~ JAMES R. KANARY, *Business Unit Manager, Healthcare Information Technology industry*

"Prior to having my own small business, I was privileged to be part of some very successful operations in small, medium, and large size businesses. Babette's comprehensive book has made me reflect on what made these operations unique and prosperous, and she has well-defined the M.O. that was involved. Her candid comments nail the realities of corporate human nature, and she provides a Progress Thought, seemingly for every day of the week. I would call *Do* YOU *Mean Business?* 'very aggressive common sense.'"

~ DAVE MARTIN, *President, Allied Window, Inc.*

"I have worked in engineering for 16 years, contributing to many internal and external multi-disciplinary team meetings. I know how important it is to have a sharing attitude towards others, and to collaborate in order to resolve differences and nurture innovative outcomes. Babette Ten Haken's book promises to be a powerful tool to help all those involved in engineering, management and marketing come together and make a measurable difference to your business. I have conversed with Babette on such topics for over 12 months now, and find her to be very passionate about her business. She is informative and dedicated to spreading great advice to help businesses like mine."

- GLEN COOPER, *Structural Engineer and Director, Avatar Engineering*

"If you're a technical professional who's involved in the sales process, you're under pressure to make pitches that convince prospects to do business with your firm. But for some reason, what you're told to do just doesn't feel right... To be successful today requires a major rethinking of what works... In *Do YOU Mean Business?*, Babette Ten Haken challenges traditional stereotypes and shows you what actually works in today's business environment."

- JILL KONRATH, *business strategist and author of* Selling to BIG Companies *and* SNAP Selling: Speed Up Sales and Win More Business with Today's Frazzled Customers

DO YOU MEAN BUSINESS?

Technical/Non-Technical Collaboration, Business Development, and YOU

BABETTE N. TEN HAKEN

Spinner Press

*Do YOU Mean Business? Technical/Non-Technical Collaboration,
Business Development and YOU* ©2012 by Spinner Press, LLC
All rights reserved. No part of this book may be used or reproduced in any manner whatsoever without written permission from the author except in the case of brief quotations embodied in articles and reviews.

Spinner Press, LLC
P. O Box 130045
Ann Arbor, MI 48113
publisher@spinnerpress.com
www.spinnerpress.com

ISBN: 978-0-9848986-5-7
Library of Congress Control Number: 2012901763

Printed in the United States of America
Cover and Internal Design by: www.Cyanotype.ca
Editing by Marlene Oulton and Gwen Hoffnagle
First Printing: 2012

Legal Considerations

The scanning, uploading, and distribution of this book via the Internet or via any other means without the permission of the publisher is illegal and punishable by law. Please purchase only authorized electronic editions and do not participate in or encourage electronic piracy of copyrightable materials. Your support of the author's rights is appreciated.

For informational purposes only, this book contains the names and web addresses of web sites created, maintained, and controlled by other individuals or organizations. The names of these organizations, and the content, images and logos of the web sites mentioned in this book, may be protected by trademark and copyright, as well as other laws, and are the property of their respective owners. Sales Aerobics for Engineers®, LLC has no responsibility and assumes no liability of any nature for the content of any web site that is mentioned in this book other than those sites which are the property of and maintained by Sales Aerobics for Engineers®, LLC. The web addresses were verified prior to the printing of this book edition. The content of these sites may change. By reading this book and/or by visiting any of the web sites mentioned in this book, you assume any and all risk including any business risk. The author and publisher of this book, Sales Aerobics for Engineers®, LLC, Spinner Press, LLC, and Babette N. Ten Haken, are not connected to, nor endorse the present or future content of any of these sites other than those sites created for and maintained by Sales Aerobics for Engineers®, LLC.

For Randy

CONTENTS

Acknowledgments	xvii
Foreword by Jill Konrath	xxi
Introduction	1
Overview – Do *YOU* Mean Business?	5

PART ONE – US VERSUS THEM: THE ELEPHANT IN THE ROOM

CHAPTER 1 – LEAVE YOUR BAGGAGE AND BIASES AT THE DOOR	13
Understand What's Holding You Back	13
Battling Silos and the Status Quo	15
The No-Silo Business Development Zone	17
Review of Main Points	19
Sales-Engineering Interface™ Tool #1: Safari Time!	20
CHAPTER 2 – WHY PEOPLE SAY NO AND STUFF GETS STALLED	21
It's Easier To Say No	21
The Anatomy of *No*	22
How Does the Interface between Sales and Engineering Contribute to *No*?	25
Getting Unstuck from the Status Quo	29
Review Of Main Points	31
Sales-Engineering Interface™ Tool #2: The Anatomy Of No	32
CHAPTER 3 – FINDING THE COMMON DENOMINATORS	33
It's about Revenue Generation	33
Common Denominators Provide Opportunities	35
The Typical Status-Quo Business Meeting	36
Put Yourself in the Shoes of Others To Establish Your Common Denominators	38
Adopt a Streamlined Approach to Communication	39

Review Of Main Points	42
Sales-Engineering Interface™ Tool #3: Finding Your Common Denominator In A Collaborative Colleague	43

CHAPTER 4 – BUSINESS BABBLE AND TECHNO-SPEAK CREATE BARRIERS — 45

Stop Lecturing and Start Listening	45
Does Slinging the Lingo Define You as a Professional?	46
What's in a Word?	48
Use Your Value Proposition To Move beyond the Status Quo	49
Review Of Main Points	53
Sales-Engineering Interface™ Tool #4: Creating Your Value Proposition By Understanding Your Core Capabilities	54

PART TWO – UNDERSTAND YOUR VALUE

CHAPTER 5 – ADAPT, ADOPT, APPLY — 57

Your Functional Role May Be Different from Your Job Title	57
It's All about Your Core Capabilities	59
Your Job Functionality Is Earned	61
Define and Use Your Core Capabilities	62
Review Of Main Points	66
Sales-Engineering Interface™ Tool #5: Determining Your Value	67

CHAPTER 6 – ARE YOU AN ORDER-TAKER OR AN INNOVATOR? — 69

Everyone's an Engineer	69
Understand and Use Your Professional Currency	70
Would You Do Business with Yourself?	73
Do You Know How To Play Well with the Other Children?	76
Review Of Main Points	78
Sales-Engineering Interface™ Tool #6: Developing And Using Your Professional Currency	79

CHAPTER 7 – WHY SOFT SKILLS ARE POWERFUL	81
What They Didn't Teach You in Engineering or Business School	81
None of Us Took Classes about How To Be People	82
You Will Be Evaluated on Your Communication Skills	84
Are Your Customers Comfortable Doing Business with You?	86
Review Of Main Points	88
Sales-Engineering Interface™ Tool #7: Words Are Powerful	89
CHAPTER 8 – WHY DO YOU WORK FOR OTHER PEOPLE?	91
It's Your Script, Not Theirs	91
Business Development Needs To Be Part of Your Formula	92
Flat-World Choices May Offer Many Opportunities	93
Do You Know How To Be the Boss?	95
Take Your Current Functional Role and Run with It	96
Review Of Main Points	98
Sales-Engineering Interface™ Tool #8: Align, Align, Align	99

PART THREE – USING YOUR PROFESSIONAL CURRENCY TO DRIVE THEIRS

CHAPTER 9 – EVERYONE IS A CUSTOMER OF EVERYONE ELSE	103
Who Are Your Customers?	103
You Are the CEO of Your Core Values	104
You Are the CEO of You and Your Career	105
Learn about the Big Picture: The Business Plan	106
Share Your Information, Because Nobody Can Deliver It Like You Can	107
Review Of Main Points	108
Sales-Engineering Interface™ Tool #9: You Are Your Own CEO	109
CHAPTER 10 – BUSINESS PLAN 101	111
Not Everyone Has a Business Plan, But They Need One	111
Why a Business Plan Is Not a Strategic Plan	113

Why a Business Plan Is Not a Matter of Inputting Facts and Figures into a Templated Document		114
Standard Elements of a Business Plan		115
Section 1:	Executive Summary	116
	Business Overview	117
	Success Factors	117
	Financial Plan	118
Section 2:	Company Overview	118
Section 3:	Status of Work	120
Review Of Main Points		121
Sales-Engineering Interface™ Tool #10: Thinking About Your Business Plan		122

CHAPTER 11 - YOUR INDUSTRY, YOUR COMPETITORS, AND YOUR MARKETS   123

Section 4:	Industry Analysis	123
	Market Overview	124
	Relevant Market Size	124
Section 5:	Customer Analysis	124
	Target Customers	125
	Customer Needs	126
Section 6:	Competitive Analysis	126
	Your Competitors	127
	Direct and Indirect Competitors	127
	Competitive Advantage	128
Section 7:	Marketing and Sales Plan	129
	Marketing Communications Plan	130
	Sales Plan and Revenue Generation	131
	Pricing Structure	132
Review Of Main Points		133
Sales-Engineering Interface™ Tool #11: Your Industry, Your Competitors, And Your Markets		134

CHAPTER 12 – ACHIEVING YOUR MILESTONES	135
Section 8: Operations Plan and Key Operational Processes	135
Day-to-Day Operational Plan	135
Section 9: Business Model	137
Does Your Current Business Model Reflect the Status Quo?	138
Key Human Assets	138
The Value of Board Members	139
Section 10: Milestones	139
Section 11: Revenue Model	140
Financial Highlights	142
Review Of Main Points	144
Sales-Engineering Interface™ Tool #12: Comparing	
Fantasy With Reality	145

PART FOUR – BECOMING THE GO-TO PERSON

CHAPTER 13 – DO YOU SIMULTANEOUSLY SIT ON BOTH SIDES OF THE TABLE?	149
Put Yourself in Your Customer's Shoes before You Try To Sell Them Anything	149
Compare the Status Quo and Their Perception of Your Value	150
Clients Avoid Making Decisions	152
The Sales Process Is Broken	153
Develop A-List Customers by Being on Their Go-To List	154
Review Of Main Points	156
Sales-Engineering Interface™ Tool #13: Be The Best And Work With The Best	157
CHAPTER 14 – THE ART OF BUSINESS DEVELOPMENT	159
Did *You* Ever Learn How To Develop Business?	159
Are Sales Funnels Another Version of Turning Your Company into an RFP Mill?	160

Have C-level Discussions with Your Customers	162
Okay, So I Have To Develop Business. Now What?	166
Review Of Main Points	169
Sales-Engineering Interface™ Tool #14: Get Out There And Do It!	170

CHAPTER 15 – YOU ARE THE PHYSICAL EMBODIMENT OF YOUR COMPANY

	171
Personal Branding Isn't Just for Them – It's for You, Too	171
Personal Branding and Technical Professionals	173
Articulating Your Core Values Online	174
The Four Attributes of Personal Branding	174
Your Personal Brand Is Your Business Plan	176
You Receive as Much as You Put into Personal Branding	176
Starting Social Networking	180
LinkedIn	180
Twitter	183
Facebook	184
Blogging	185
Review Of Main Points	187
Sales-Engineering Interface™ Tool #15: Your Personal Brand	188

CHAPTER 16 – TAKE CONTROL OF THE CUSTOMER ACQUISITION AND BUSINESS DEVELOPMENT PROCESS

	189
Business Development Is a Process of Discipline and Alignment	189
Leverage What You Know against What You Don't Know	190
Do Not Expect Your Clients To Think in a Straight Line	192
Understand Where Your Customers and Prospects Go	193
Review Of Main Points	196
Sales-Engineering Interface™ Tool #16: Do I Really Know My Customers?	197

CHAPTER 17 – PUTTING IT ALL TOGETHER	199
Business Development for Technical and Non-Technical Professionals	199
Are Technical Professionals the Stewards of the New Business Development Paradigm?	200
Bring Your Personal Core Values into Every Professional Interchange	201
It Won't Be a Piece of Cake, But If It Were, It Wouldn't Be Fun	202
Staying Focused	205
About The Author	207
Resources	209
Blogs	209
Sources Of Quotes And Books Cited, By Chapter	211
Glossary Of Terms, By Chapter	215

ACKNOWLEDGEMENTS

I was close to completing this book's manuscript when a young businesswoman I was coaching asked me to share an excerpt with her. I read her the opening paragraphs of the first chapter and she exclaimed, "Babette, how can you know this? Do you work at my company?" "Yes," was my calm reply. "I've worked with many companies just like yours. I've walked in your shoes many times."

This book sums up my professional experiences over the past twenty-five years. I've had the opportunity to work with colleagues and mentors from a diverse mix of educational institutions, corporations, and industries across the globe. I am grateful not only to these entities, but also to the incredible friends I met along the way. They inspired insight. They encouraged me to always do my best work and never give in to self-doubt or second-hand opinions.

First and foremost, I want to thank Jill Konrath who, since 2005, has kept my business head on top of my shoulders and focused. Jill helped me find my blogging voice. She has been both a guiding light and a no-nonsense sounding board for me. Her books are my sales bibles. My thanks, as well, to Katie Konrath, my technical and blogging coach for the first version of my Sales Aerobics for Engineers® blog. After critiquing the first few blog posts, she told me to take off the training wheels and keep blogging. And I did!

I have profound gratitude for the technical and non-technical professionals who form my client base, and my LinkedIn, Twitter, and Facebook colleagues. Their insights in response to my blog posts, comments on

LinkedIn discussion boards, and our in-person and online interchanges fuel my own aha! moments. We all grow as individuals and business professionals because of our desire to collaborate.

I am thankful that Matt Barcus and Carol Metzner, of CivilEngineeringCentral.com, gave me an almost immediate opportunity to guest blog for their award-winning civil engineering site back in 2009. They are tremendous barometers of the industry and have been patient, sympathetic, and downright funny whenever I hit a creative block.

I would not have become as deeply involved in business development in the manufacturing sector were it not for the advice and opportunities offered by the veteran sales professionals I met and worked with along the way. I carry a pair of safety glasses in my car as standard operating procedure, thanks to them. I love manufacturing equipment and remain absolutely fascinated by the engineering design that goes into such amazing pieces of integrated technology. The manufacturing and technical service sectors remain the backbone of every country; it is up to technical and non-technical professionals to collaborate on their behalf. It is my distinct honor to serve these companies' business development needs.

Thanks to an initial editorial recommendation from Anthony Fasano, P.E., I went on to assemble a "dream team" to help me get this book out of my brain and into digital and paper format. Kudos to my editor, Marlene Oulton, whose experience and outrageous sense of humor kept me moving forward. Marlene is the only person I know who can identify via email when your eyes are bugging out and you are completely overwhelmed and heading towards your mental ledge. A huge thank you to my hero of a webmaster, Doug Yuen, who created my Sales Aerobics for Engineers® blog and website, and the website for this book. A new personal blog at www.babettetenhaken.com is on the drawing board. Doug is incredibly patient and an amazing source of all things Internet. Thank you to my virtual assistant Jefre Keep, whose humor and behind-the-scenes artistry make our collaboration click. This book would never have become a reality if it were not for the publishing expertise of Lynne Klippel and the amazing graphic design work of Sarah Barrie. And last, but certainly not least, I am grateful for the professional proofreading expertise of Gwen Hoffnagle. I had forgotten what the English language really looks and reads like.

We find people along the way who intersect with our lives and encourage us to express ourselves and confidently let our lights shine. I had the honor of meeting and sharing ideas with William Emmet Root a few weeks before his passing in April, 2011. Bill, you were and remain remarkable. Thank you.

Finally, my wholehearted gratitude to my wonderful husband, Randy, and my children Anna, Andrew, Alex, and Jill, for their patience, interest, feedback, and just plain support while I wrote, blogged, tweeted, and contemplated this book. You are the core of my soul.

FOREWORD
by Jill Konrath

The sales process is tough. If you're in sales, you know how much time it takes to set up meetings with potential prospects. They're not receptive to your advances. They'd rather stay with the status quo than change. The budgets are tight and all they're concerned about is price.

If you're a technical professional who's involved in the sales process, you're under pressure to make pitches that convince prospects to do business with your firm. But for some reason, what you're told to do just doesn't feel right.

Sound familiar? The truth is, in the past few years your prospects have changed – radically. Since virtually everything they need to know can be found online, they don't need to meet with you. Nor do they have the time. Everyone is crazy-busy, trying to handle more work and impossible deadlines with fewer resources.

As a result, their expectations of us, as sellers and technical professionals, have changed, too. They're tougher on us. More demanding. We have to prove we're a valuable resource before they'll even consider having a relationship with us. But saying good things about ourselves or our company falls on deaf ears.

Despite all this, fewer than 10 percent of sellers have altered how they approach prospective clients, create opportunities, or differentiate themselves from competitors.

To be successful today requires a major rethinking of "what works." In my first book, *Selling to BIG Companies*, I introduced new strategies to help sellers get their foot in the door of targeted accounts. In my second book, *SNAP Selling*, I focused on new strategies for dealing with frazzled, harried decision-makers.

Babette Ten Haken challenges stereotypes as well. I first met her seven years ago, when she called me with a question. Having recently taken on a sales role, she was perplexed at the divisions between the sales and technical functions. And, she felt like she was being pushed to do things that not only didn't work, but also compromised her belief system.

She was right. And since that initial conversation, she's been a woman on a mission to help sales and technical professionals be more successful with business development. In *Do YOU Mean Business?*, she challenges traditional stereotypes and shows you what actually works in today's business environment.

You'll find answers to questions such as:

- What should your sales process look and sound like when you're interfacing with prospects and current customers?
- What resources are available to you as technical and non-technical professionals working together?
- How can you become valuable resources to your customer's decision making?

If you knew more about Babette's background, you'd realize just how much she knows where you're coming from. Trained as a scientist, she spent years facilitating left brain-right brain meetings as a marketing research professional in the pharmaceutical industry. Following that, she transitioned into total quality management and Voice of the Customer research.

To her, the cross-over interface between sales, business development, and engineering is fluid. For over 25 years, she's been doing this "simultaneous translation" between technical and non-technical colleagues that resulted in very productive and profitable outcomes.

When you read *Do YOU Mean Business?* you'll see what I mean. She'll shake up your perceptions and then deftly guide you through what it takes to be successful. It's well worth your time to read it.

~ Jill Konrath, business strategist and author, author of *Selling to BIG Companies* and *SNAP Selling: Speed Up Sales and Win More Business with Today's Frazzled Customers*

INTRODUCTION

I've always had a knack for getting people to say what they mean in a way that is understood by folks sitting around the table. Maybe that's because I find the similarities between people far more fascinating than the differences. That is why I chose a dual major in college – physical anthropology (the high-level, strategic "why" context of systems) and evolutionary genetics (the tactical "how" the system is implemented in response to the environment).

My first job was in the pharmaceutical industry, as a newbie clinical research associate. Again, I was correlating "how" diseases manifest themselves within the "why" context of populations. In order to gain perspective, I realized I needed to take a step back from the issues at hand in order to establish an even broader context for my observations. I started to ask how what I was doing as a researcher fit into the goals of the organization I worked for and the trends emerging in the pharmaceutical industry. My management became aware of the enhanced perspective I was now bringing to meetings. In researching answers, I was collaborating with a lot of colleagues across disciplines and professional organizations.

I became the go-to girl for my company. I was called into meetings for *simultaneous translation* between technical and non-technical professionals, research and development, and marketing/sales. I facilitated meetings, asking participants to tie their findings to the bottom line and strategic goals of the company, as well as to industry trends. Discussions became productive instead of adversarial. The teams I worked with became profitable. What I unintentionally fell into, due to management recognizing the potential of my skill set, became the fulcrum of my career. As a result of this new career direction, I obtained formal training in Voice of the Customer methodology and facilitation, market research facilitation, and Six Sigma methodology, so that I could put these skills to work for my clients.

I realize that no matter what my current job title or function, I consistently bring this simultaneous-translation skill set to the table. It is my common denominator across whatever I am asked to do on behalf of my clients. This capability is the value I bring to my company, my clients, and myself. I have always incorporated this communication style into my business development, facilitation, and coaching strategies. This method works for all facets of business, whether in the business-to-business (B2B) or business-to-consumer (B2C) sectors. It involves adapting one's perspective and adopting a cross-functional communication style. It takes no time at all to apply this strategy to achieve results.

As I continue to work with the B2B sector, I realize that people do not take the time to evaluate what their colleagues do in their companies. And many do not fully understand the value they themselves provide to their organizations or how their capabilities can be fun, synergistic, and drive revenue. There is a cross-functional disconnect fueled by our professional disciplines.

They aren't the only ones grasping at these professional straws. No matter how young or old my clients are, or whether their businesses are mature or start-ups, their understanding of the business development process remains weak. Even very smart technical professionals and business owners have difficulty articulating the value their product or service provides to their customers. They have a poorly defined concept, if any, of potential markets. Their idea of what constitutes a sales pitch often leaves potential investors and clients confused rather than excited.

When some of my clients downsized in 2008, owners asked their engineers to start generating business. These technical professionals were very uncomfortable in their new roles, and they rattled off one-size-fits-all sales spiels. This was not only inadequate, but also made them look out of touch and amateurish to prospective clients and current customers. Yet their sales pitches represented their perception of business development! If these companies could not demonstrate the value of doing business with them to their current clients, how could they expect to win business from prospective clients?

This disconnect wasn't confined to my client base, either. I contribute my skills pro bono to several local business incubators. I thought that at least the young entrepreneurs I worked with would exemplify an updated, enlightened, cross-functional mindset. Yet many of these very bright young people naively perceived that investment dollars "happened" as a reward for a great idea! And as a reviewer and judge of business plans for business competitions, I found the same cross-functional disconnect alive and well. Many of the business plans I read lacked the logic and documentation to tell the company's story and provide a compelling reason to award funding.

New MBAs and engineering graduates I coached communicated their frustrations to me regarding their current professional roles. These very bright graduates described cross-functional meetings with engineering and marketing/sales teams that were not going productively. Each side felt their discipline drove the other side's output. These talented young professionals didn't have the right tools for establishing a dialogue of compromise – let alone collaboration – with their colleagues.

These are the same issues I encountered twenty-five years ago when I was the corporate newbie, and the same dynamics I have experienced throughout my career. Have these scenarios really not changed at all? Clearly the professional disconnect is alive and well. I am still spending more time with my clients untangling the misconceptions of discipline-driven, status-quo, and the-way-things-are mindsets than I am working on pointing their companies towards the way things *should* be. Something has to change!

My clients were so stuck in the way things are that they did not have a clue how to accomplish the simultaneous translation that I was able to achieve for them. As I taught clients and students to think cross-functionally, they became less frustrated with each other and more successful at working as a team to win new business. They would have continued to spin their wheels in their own status-quo mindsets had I not provided them with the tools to move just one millimeter outside of their comfort level.

I had to be able to teach my clients to sit on both sides of the technical/non-technical table and exchange ideas effectively, productively, and profitably. The economic meltdown of 2008 is redefining the business development paradigm. Too many professionals are getting left behind or mired in status-quo environments, even in well-known companies. I felt I had to reach out to others besides my own client base.

This book will teach you what I have learned. Business babble and techno-speak create barriers to collaborative business development. There's no reason to perpetuate the entrenched, discipline-driven, status-quo behaviors preventing your company and you from moving forward. And you can't move forward unless you understand what is holding you back. This book gives you that 10,000-foot eagle's-eye view of the business development landscape so you can develop the mindset and communication skill set to increase the value you provide to yourself, your clients, and your organization.

OVERVIEW:
Do *YOU* Mean Business?

Are you the:

- *Owner/President/CEO of your company?* How well do you recognize and define a business opportunity for your technical and non-technical staff?
- *Vice-President of Sales, Engineering, or Business Development?* If you are winning business by trying to be all things to all people rather than focusing on core capabilities, you may be creating production backlogs and lost profitability.
- *Sales-Engineer?* Are you comfortable assuming a business development role or is something lost in your translation?
- *Recent technical or non-technical graduate filling your first professional role?* Do you feel limited in your current position and want to be more to your organization?
- *Entrepreneurial start-up?* Do you have the tools to know what decisions need to be made, at what time, and by whom? (Hint: the answer to all three starts with *you!*)

How capable of and comfortable with participating in today's business development continuum are you? Can you put yourself in your customer's shoes? Do you understand that perpetuating the what-got-you-there mindset will not carry you forward anymore? *Do* YOU *Mean Business?* shows you how to develop the perspective and skills that will differentiate you from your competitors and colleagues. Wouldn't you like to make yourself a valuable part of the business and revenue development cycles for your company?

PART ONE –
Us Versus Them: The Elephant In The Room

Part One of this book gets the elephant out of the room once and for all. It identifies the signs and symptoms of the us-versus-them mentality and why it doesn't work anymore. In fact, it never worked at all. You will learn to:

- Identify the biases and baggage that create the status quo, hold you back, and prevent you and your company from moving forward
- Understand why *NO* is a powerful and controlling word in companies, and how you can anticipate and diffuse its use in your organization
- Find the common denominators across disciplines that break down barriers and dispel the status quo
- Communicate collaboratively and productively without relying on the business babble and techno-speak that reinforce siloed mindsets and discipline-driven, status-quo infrastructures

PART TWO –
Understand Your Value

In Part Two, you define your core capabilities so you can build your value based on the common denominators you bring to all aspects of your profession. You will learn to:

- Understand why your functional role may be different from your job title, and why that is a good thing
- Determine whether you are an order-taker/implementer or an innovator, and how to build your core capabilities around what's natural for you
- Become comfortable growing and using your not-so-soft skills instead of hiding in your cubicle
- Determine why and whether or not it makes sense for you to work for other people

PART THREE –
Using Your Professional Currency To Drive Theirs

Part Three involves developing your understanding of the dynamics and processes of business development from both technical and non-technical perspectives. You will learn to:

- View everyone as a customer of everyone else
- Assume the role of CEO of you and your career, not just your job
- Understand the importance of the elements of a business plan
- Start thinking like a business owner in order to bring that perspective to your career functionality

PART FOUR –
Becoming The Go-To Person

Part Four allows you to start shifting your professional gears by incorporating your simultaneous-translation perspective and skill set into your discipline. You will learn:

- How to develop business and understand the dynamics of a professional conversation by simultaneously sitting on both sides of the table
- Why it is important that your customers see you as the physical embodiment of your company and your personal brand
- The dynamics of how your customers make decisions
- Why putting it all together allows you to provide greater value to yourself, your company, and your clients. You are, after all, a wonderful and continual work in progress

The Utility Of This Book

This book builds on itself. It is meant, initially, to be read in the order written. Once you have read the book, you'll know which subjects will take work and which are naturals for you. Ultimately, *Do* YOU *Mean Business?* is meant to be the book you turn to as a resource, a jolt in the arm, and a kick in the pants for your professional development and your day-to-day business interactions. Return to the Sales-Engineering Interface™ toolkit exercises, which appear at the end of each chapter, as often as needed to guide your forward progress in technical/non-technical collaboration.

After you have read the book, I recommend you start defining your core capabilities. Once you've identified your core capabilities, take them into your next meeting and refine them as you continue to observe and participate. Return to this book to refresh your memory.

Breaking away from the linguistics of your professional discipline will be one of the most important outcomes of reading this book. At your next meeting, remain engaged by analyzing the dynamics of the functional roles of everyone seated around the table. Understand which attendees will be most affected by the new perspective you now bring to the table. Don't be afraid to address specific comments to specific individuals, providing you have done your homework! Take note of responses and questions asked by others in the meeting. By developing the ability to simultaneously translate – and articulate – expectations, processes, and deliverables from technical and non-technical perspectives, you will have developed another valuable core capability. Then prepare for subsequent meetings by incorporating the perspective you gain from Part Three, which addresses the fundamentals of the business planning process. You'll not only get more out of these meetings, but you will contribute more as well.

Be patient with yourself!

Don't over-analyze, and don't try to cram everything you learn into your next meeting. You will be overwhelmed with this information initially. Build your knowledge base and comfort level incrementally. Return to this book when you feel unsure or uncomfortable. Use it as a resource so you can ask better questions, speak with confidence to colleagues and clients, and increase your knowledge base. Remember, you are a work in progress. This book will allow you to develop new skills that you can layer onto your existing set of core capabilities.

The Sales-Engineering Interface™ Tools at the end of each chapter serve as a quick reference for techniques and tips to employ during meetings, presentations, and conference calls. In the back of this book, I provide some books, websites, and other resources, as well as a glossary of terms. You may find these helpful as you develop greater awareness and interest in your growing business development role.

Do YOU *Mean Business?* is written as a catalyst for you to liberate yourself from your own status quo and the mindset of your professional discipline. It is my sincere desire to provide you with insights and practical tips and tools to help you increase your perspective outside of your discipline, provide enhanced professional value to your associates, and become a resource for others.

Enjoy your understanding of how *you do* mean business for your company!

Part One

US VERSUS THEM:
THE ELEPHANT IN THE ROOM

LET'S TALK ABOUT
THE ELEPHANT IN THE ROOM

CHAPTER ONE

LEAVE YOUR BAGGAGE AND BIASES AT THE DOOR

Understand What's Holding You Back

It starts at the Monday morning meeting and ends Friday afternoon, if it ends at all. You probably take it home with you. The elephant in the room, that is. It's the subject that nobody wants to discuss, and the elephant is everywhere. Your company has elephants in various rooms, your clients have elephants in their rooms, and your institutions have elephants in classrooms. It's that us-versus-them mentality. Technical professionals versus non-technical professionals, techies versus non-techies, geeks and nerds versus motor-mouth biz and sales types – these are just a few of the phrases used to describe those elephants in many organizations.

The us-versus-them mentality doesn't work. In fact it never has. If you poll your colleagues about the best teams they ever worked on and the best projects they ever completed, you will hear stories about groups of people who were willing to exchange ideas across disciplines and explore all available options. The "dream team" groups they describe were willing to ask questions and learn the language of the disciplines in which they needed to communicate. Your colleagues will also tell you that these teams broke up after successful product launches and were not asked to re-assemble again to repeat their formulas for successful innovation.

Why not? Their success was regarded as a fluke. But it wasn't. They knew they had to kick the elephant out of the room in order to deliver on the project goal. And kick it out of the room they did! Frustrated that their group synergy wasn't recognized or rewarded upon project completion, many team members subsequently left the company in search of a position at a company that encouraged collaborative and productive dynamics. Other team members elected to remain at the company to fulfill their tenure, yet the memory of their dream team remains as one of the best things they ever did professionally. They inspired one another. Success is success, and it happens for distinctive and quantifiable reasons.

The reality of most work environments can hardly be described as a dream situation. How much of your workday is spent bickering with and validating yourself to the us-versus-them mentality that exists *within* disciplines as well as *between* disciplines? These less-than-inspiring dynamics create busywork and they don't contribute to developing business. Busywork and the dedication companies devote to its preservation create obstacles and excuses for us to underperform and remain unengaged with our colleagues, and bring neither the attitude nor the energy to our professions that we know we have within us. The workplace can degenerate into a lot of finger-pointing. And let's face it, if everyone is blaming everyone else, then management can avoid addressing the underlying root causes of the disconnect!

The us-versus-them mindsets in our workplaces hold us back instead of moving us forward. After all, when we sweep away the dysfunction of most business models and the workplaces they create, our professions really *are* about generating business, whether this responsibility is specifically stated in our job descriptions or not. You *do* need to play a role in contributing to your company's bottom line. You *don't* need to constantly role-play in reinforcing your company's status quo. Being a technical professional does not absolve you of your role in acquiring and preserving customers. Being a salesperson does not allow you to use short-term, give-it-away sales methods rather than proposing technically realistic, production-feasible, and cost-effective solutions for your clients. This book provides you with a 10,000-foot eagle's-eye view of the situation so you can participate in the dynamics of the business development process, even if you are a technical professional! You *do* mean business, don't you?

Developing the ability to put your job functionality into a broader business context is an essential element of moving beyond the us-versus-them mindset. Whether you are a technical professional or a sales and marketing professional, what you do impacts not only the strategic direction of your company, but revenue generation as well. The Monday morning workplace will become less of a soap opera and more of a business development forum if you listen and learn from an objective and informed perspective rather than from a personal, discipline-specific one. After all, root causes – and the us-versus-them mentality – can have really large contexts. This revelation begins once you decide to adapt, adopt, and apply business and technical information to your job responsibilities.

Battling Silos and the Status Quo

Traditional business models create divisional or departmental silos. This structure typically restricts the flow of knowledge between departments and individual resources. Everyone is loyally protective of their divisional and departmental knowledge cache. This type of business model creates a lot of control…and a lot of elephants in the room as well. The lines are demarcated by discipline, division, mindset, training, and just about any other type of difference that can be identified within your organization. These traditional attitudes are the status quo, or "the way it's always been done" of your business model. They hold up the forward progress of cross-functional, collaborative business development strategies.

Who likes to go to their place of work and feel like they are entering a battleground? Not anyone I know! Yet many of us treat our careers and our daily responsibilities as if we have to be continuously vigilant and on guard. We are put on alert, defending our jobs instead of our market share, battling our colleagues instead of our competitors. We have sales quotas to meet and non-billable project hours to reduce. Management is looking over our shoulders and talking cutbacks and downsizing. Sometimes the management looking over our shoulders is our parents who own the company we work for! We often feel our company has a ready-fire-aim business development strategy. We are put on the defensive from the minute we walk through that door each Monday morning. It's exhausting and uninspiring!

Take a 10,000-foot eagle's-eye view of this battlefield and its root causes. You and your colleagues form, in a sense, a special forces unit for revenue generation for your company. That's what you all were hired to do, whether it states that in your respective job descriptions or not. In line with this analogy, wouldn't you want your buddies in your unit to have your back, just as you would have theirs? Your perspective should be external, looking towards the battlefield. Yet, as you experience on a daily basis, the in-fighting never seems to end. Instead of you and your colleagues working like a well-drilled combat unit, everyone is perpetually dragging their respective elephants into the room to engage in minor skirmishes and in-house turf wars. Who is going to win a battle with so many pachyderms littering the field of engagement?

Marketing timelines and methodologies (qualitative, quantitative, subjective, and extrapolative) seem to be counter-intuitive to the analyses required by technology, the hypothetico-deductive scientific method, and the Define – Measure – Analyze – Improve – Control (DMAIC) approach used to eliminate causes of defects. Marketing and sales may want quick responses to market demands. Research and development (R&D) deliverables may be limited to line extensions due to old equipment and lack of revenue to invest in new machinery. Engineers will hold their technical cards close to their chests, and marketing will complain about their engineers' lack of acceptance of promotional strategies. Everyone retreats back into their respective departments, wounded in another failed cross-functional meeting, and the status quo persists. This scenario, played out in countless companies on a daily basis, is a major reason why companies stall. How can you have each other's backs if you are at each other's throats?

Business development is a dynamic playing field. The perspective you must take is larger than the next design project, marketing campaign, or signed contract. Success involves the efforts of everyone, simultaneously pulling together.

The No-Silo Business Development Zone

Keith Sawyer, in his book *Group Genius: The Creative Power of Collaboration*, uses the business case of W. L. Gore & Associates to demonstrate what a collaborative business environment can achieve. The company invented GORE-TEX® waterproof material, among many other commercialized and profitable innovations. In describing their workplace, which consists of self-managed teams that assemble, re-assemble, and re-configure as needed, one employee describes: "Your team is your boss, because you don't want to let them down...Everyone's your boss, and no one's your boss." (*Group Genius: The Creative Power of Collaboration,* p.18.) This company and these individuals have suspended titles, disciplines, and biases to work together collaboratively for innovation and revenue generation.

Does the business model for W. L. Gore & Associates seem like light-years from your workplace? Developing this type of business model didn't happen for them overnight. Gradually, however, and with the wisdom of upper-level management, the company elected to restructure their business model and assemble the types of professionals who work best in this seemingly free-form, collaborative, innovative atmosphere. Although this structure may seem overwhelming to you, what would happen to your company's revenue stream – and to you professionally – if you were allowed the freedom to flex your collaborative muscles a bit more in the workplace?

The status quo is the most comfortable mode of operation for management. Any decision is perceived as change. And when all departments are set up to reinforce status-quo thinking, any decision to change is perceived as being disruptive! "The way things have always been done" is seen as the reason the company has gotten to where it is today. Change involves moving one millimeter outside of management's collective comfort level. And who wants to rock that boat? *Making Smart Decisions*, published by the Harvard Business Review, discusses, among other topics, how organizations gravitate towards maintaining the status quo. They make decisions to keep things the same, even if the status quo doesn't truly serve their long-term goals. How many companies engage in avoiding change rather than adapting a robust and responsive competitive strategy that challenges their talents and core capabilities?

When I begin working with a client, regardless of their annual gross revenue, number of employees, or length of time in the marketplace, the client and I identify what the status quo looks like in their organization. The company has always looked a certain way, been organized a certain way, employed people with specific, traditional job titles, and everyone has been expected to interact with others in a traditional manner. There is no coloring outside these very homogeneous lines! Many entrepreneurs have come from traditionally organized companies. Even these entrepreneurs, intentionally or not, bring the baggage of status-quo mindsets and prior business models into their entrepreneurial plans. They end up dragging the elephants from their former companies into the room and reinventing the same wheel that they thought they were leaving behind!

It's time for everyone to pry our brains off status-quo mindsets and the elephants in the room that they create and perpetuate. Companies and disciplines hold on to status-quo mindsets for dear life because the alternative – change, the great unknown – appears uncomfortable and risky. What would happen if you placed a sign at the entrance of your company that stated: "Leave Your Elephants at the Door"? It's up to you to bring that attitude into your professional space each day. Take personal responsibility for liberating yourself from status-quo thinking so you can provide value to yourself, your organization, and your customers. Shall we begin?

REVIEW OF MAIN POINTS

1. The us-versus-them mentality doesn't work. Professionals waste a lot of time justifying their disciplines to each other instead of focusing on customers and revenue generation. Understand where, why, and how this perspective is manifested in your workplace.

2. Make sure you have each other's backs. The workplace can become a battleground of infighting instead of a globally competitive playing field. Take a 10,000-foot eagle's-eye view of your project and deliverables. Take the entire team to the finish line, together.

3. Preserving the status quo is limiting. Change is perceived as disruptive and risky. Individuals, institutions, groups, organizations, and cultures would rather preserve and control the status quo – the way things have always been done – than move one millimeter outside of their comfort levels.

4. While you may not be able to re-organize your company, you *can* re-organize yourself. Modifying how you communicate and implement your skill set as a business or technical professional can have meaningful and profound effects on project outcomes and revenue generation.

SALES-ENGINEERING INTERFACE™ TOOL #1: SAFARI TIME!

Yep, you got it! It's time to locate the elephants. All of them. As you move through your workday and workweek:

1. Start a list of all the rooms in which you've located elephants. Which departments have the most elephants? If you could name the elephants in terms of the mindsets they represent, what would they be called?

2. If the elephants of each internal department of your company could talk to each other, what would they be saying?

3. Apply this exercise to the customers or academic institutions you deal with. They have elephants, too. Identify and characterize them. What are their elephants saying to your company's elephants?

4. What elephant(s) do you personally carry into the workplace each day? Which one(s) do you take home with you each evening?

CHAPTER TWO

WHY PEOPLE SAY NO AND STUFF GETS STALLED

It's Easier To Say No

"Naysayer: Someone who systematically obstructs some action that others want to take."
~ www.thefreedictionary.com

No is a powerful word, and the us-versus-them mentality can involve a lot of "no" and "you-can't-do-that" tactics. This behavior pattern simultaneously reinforces the status quo while perpetuating the us-versus-them mindset! Whenever there are decisions to be made, naysayers will come out of the woodwork and seat themselves at the conference room table, bringing their elephants with them. Sometimes entire divisions become known as the "No, you can't do that" people. There's usually some history surrounding divisions, departments, and individuals known for saying "you can't do that." Being a naysayer serves a very useful function in companies with traditional, siloed infrastructures.

While I have watched entire departments become obsessively involved in turf wars and being naysayers, I didn't get stuck there, and neither should you. However, understanding the history and context of how *no* is used in your company to control and stall is important information to have. And while I could devote an entire book to various ways in which people and departments sabotage each other with *no*, it's not the best use of anyone's time, and certainly not the focus of this book. It's more productive to learn how to adopt that 10,000-foot eagle's-eye perspective in analyzing how and why *no* is used by your company and your customers. Understanding the context of their entrenched positions, rather than reacting to or even participating in them, enables you to move towards a more productive outcome.

Let's face it; it's just plain easier to say, "No, you can't do that" or "No, we won't buy your solution" than saying, "Yes, we will." The complexity of interpersonal relationships in the workplace and the competitive marketplace compound why people are looking for reasons *not* to make decisions. Making a decision involves hard work, analysis, team-building, consensus-building, and moving away from what's comfortable. That's risky business for many people. The anatomy of *no* starts with the understanding that decision-makers are completely overwhelmed by demands on their time. It is up to you to make it easier for the decision-maker to say yes. That means doing your homework and providing data that anticipates and reflects the decision-maker's needs, not your own needs and perspective.

The Anatomy of *No*

How can you be proactive in assembling data that moves decision-makers towards a positive outcome? Start your homework by understanding the reasons why they say no more frequently than yes. There are many reasons why decision-making gets blocked or stalled. You, yourself, may have contributed to these scenarios. Perhaps you recognize some of these themes:

1. The decision-maker feels they have been blindsided by the need to say yes. In other words, the decision-maker hasn't done their own homework and is passing the blame and responsibility for their indecision on to someone else. This stall tactic usually is expressed as, "Why don't you prepare additional information so I can make the decision?" except there is never any amount of data sufficient to allow the decision to be made.

2. The decision to be made involves a solution that is met with skepticism and doubt. It is perceived as involving change and moving away from the way things are normally done. The result of the decision – change – is perceived as too pervasive and risky. The project is put on hold, usually expressed as, "We will think about it" or "Put that on next year's budget" (when the response will be the same).

3. The decision crosses emotional and actual boundaries into other people's departmental and organizational turf. It is perceived that authority is being challenged or an intentional workaround has been created to keep a serial naysayer out of the process. That individual finds out, takes things personally, becomes offended, and vetoes the project.

4. You unintentionally bring decision-makers into the process too late. You blindside them with that important decision they need to make. You apparently did not feel they were important enough to be included in the meetings leading up to their need to make the decision! Your prioritization, timeline, and seemingly eleventh-hour sense of urgency tells these decision-makers that you do not understand the types of information they require in order to make this type of decision. No is their prudent response, and your credibility has suffered in the process.

5. The decision-makers are protective of their jobs. There is past history involved. They may have taken a risk and were burned by shortsighted planning or poor execution by (possibly former) employees. These decision-makers no longer wish to play in what they perceive as a high-stakes game. Because they have become more conservative and risk-averse, they postpone, delay, or shoot down the proposed project.

6. There's a tradition of interdepartmental *no*. A push from the marketing department is a red flag for a push back by the engineering or R&D department. If a non-technical discipline justifies a project as a priority, there must be something wrong with their rationalization. "Because

the marketing department says so" is reversed to "The R&D folks say, 'No, you can't.'" And you can interchange the names of internal departments engaged in this scenario: marketing versus sales, marketing versus engineering, engineering versus operations, etc.

7. In a specific form of examples 1, 3, and 4 above, legal and financial colleagues are brought into discussions too late, after strategies have been decided. Risks have been evaluated and dismissed, solutions presented, and everyone needs money to proceed. The financial folks may be referred to as bean-counters, but engineering can't engineer, and marketing and sales can't market and sell if there is no anticipated return on investment. Finance and legal professionals are cautious; there is always the possibility of unintentional dissemination of intellectual property when the customer's engineer calls your engineer to kick around designs for a potential project that typically ends up being awarded to another vendor. What about your current customer's elongated billing cycle that burdens cash flow and profitability? Also consider the salespeople who tie up internal resources chasing sales that go nowhere. When blindsided, either intentionally or not, "No" or "Let's hold off on making this decision until we have all the facts" will be predictable answers from legal and finance professionals. Can you blame them?

If any of these scenarios sound familiar, pull yourself out of your emotional association with the departments and players and adopt a more historical, contextual perspective. You can't move forward until you understand what is holding you back. From your 10,000-foot eagle's-eye view of the situation, company history supports the wisdom of everyone's continuing to hunker down into "the way it's always been done." Otherwise, if one department changes, there might be a domino effect on the rest of the company, and the potential for disorder and chaos!

> *If all you thought you were doing was asking for a simple yes, then you had your head buried in the sand! You asked a very loaded question.*

Whether you are directly involved in the decision-making process or take a supporting functional role, think about moving yourself one millimeter outside your comfort level rather than changing your entire organization. Before "You can't do that" escapes your own lips one more time or you pull data to support that negative or risk-averse position, ask yourself *why* they can't do that. Perhaps they can. Perhaps you know of information, technology, context, or factors that can turn "You can't" into "We can."

What would meetings look and sound like if everyone seated around the table, computer screens, and/or telephones moved one millimeter outside their comfort levels and worked towards developing a we-can-do-that mentality? Saying no or stalling decisions might be replaced by enhanced teamwork that involves moving beyond the status quo. Meeting preparation might become enjoyable. Think about the net effect of everyone looking forward to meetings and proactively anticipating the types of information they could present that would promote synergy among participants.

How Does the Interface between Sales and Engineering Contribute to *No*?

How can technical and non-technical professionals move beyond discipline-driven status-quo mindsets and get on the same page so that projects aren't stalled and the sales cycle isn't derailed? This book focuses on that particular aspect of business development. I don't need to tell you that these disciplines have long histories of disconnects in the workplace. The sales function is responsible for bringing in contracts for technical departments to execute and implement. Everyone is supposed to be working together for seamless, collaborative production of solutions sold. Yet that very mandate is counterintuitive to each discipline's status-quo, stereotypic perspective of each other! How many of these solutions are actually delivered on time, as promised, and under budget?

Since you can't move forward until you understand what is holding you back, let's dive in and take a 10,000-foot eagle's-eye view of how "no" and "you can't do that" are used at this interface. Perhaps these scenarios sound familiar:

1. Your salespeople treat sales engineers and technical professionals like tools in a toolbox. Once the technical professionals have fulfilled their functionality within the sales cycle, they are promptly returned to their cubicles. This is the "apply-as-needed" scenario. There's a history behind this behavior pattern; it's usually based on past experience with bringing engineers into the sales cycle either too early or too late. In most siloed, division-based infrastructures, technical professionals don't have the opportunity to become familiar and comfortable with the dynamics of the sales process. Since they are perceived as a liability rather than an asset within the sales cycle, technical professionals are applied as needed, as well as inconsistently.

   Sales engineers and other engineers may have gained reputations for talking way too much and too long about all the cool technical features of the product or service being sold. What's of interest to them may not be a priority to corporate decision-makers. Since engineers are more comfortable seeking peer conversations in meetings, they may direct the majority of their conversation to the customer's engineers, and exclude the other decision-makers seated at the table. After all, if a sales engineer has been called to the meeting, then the issues must all be about the technology, right? Wrong! The sales engineers don't understand this disconnect, because they usually are not involved throughout the sales process, and this cycle is perpetuated by your company's status-quo culture.

   In addition, the salesperson may or may not have prepped the sales engineer in anticipation of this important meeting. As a result, if questions are asked of the sales engineer, they bring up issues that the salesperson has already painstakingly addressed and pre-negotiated with internal management. The sales engineer may end up telling the customer "No, we can't do that," when, in fact, your company has told your salesperson "Yes, we can." As it is, your proposed solution may be controversial or involve a lot of change. The customer may be looking for a reason to stall or prevent the decision from moving forward. Lack of communication between your sales and engineering functions causes the customer to second-guess the value of your solution. The close of the sale starts to slip away. What does preserving your company's status quo, in terms of lack of communication between sales and engineering, end up costing your company in lost sales opportunities?

2. In the "thanks-but-no-thanks" scenario, salespeople bypass engineering assistance during the sales process because a sales engineer is not available. Your company, in a move to counteract the apply-as-needed scenario, has mandated that a sales engineer must be involved in every sale. The engineering department has made the technical aspects of your products and services sound like a cosmic mystery that only technical professionals hold the answers to! (One version of "No, you can't do that" in this scenario.)

   What's the history behind this scenario? Perhaps your company does not employ enough sales engineers to go around. Maybe there are only a few sales engineers who are conversant, confident, and make good partners with salespeople. Is there a "rock star" salesperson who hogs all the resources? Does your company have a quota-driven status quo? If so, your sales engineers get sucked into working with sales reps who make appointments with unqualified prospects ("churning and burning") simply to demonstrate to management that they are out there selling.

   Regardless of the circumstances, your company feels it is all about the availability and timing of its internal technical resources in meeting with the customer's decision-makers. News flash: the customer's decision-makers' schedules take precedence over your sales engineers' availability. If the salesperson can get an available sales engineer at the meeting to close the sale, that's great. If not, the salesperson will have to wing it on their own. (Another version of the no-you-can't-do-that-to-me scenario.) The salesperson is not in a position to stall their potential client for weeks or months until a sales engineer is available; your customer may perceive that the salesperson is not competent or that the engineering department is the tail wagging the sales dog.

Timing does, indeed, become an issue when chasing decision-makers, sales quotas, and commission dollars. Salespeople may not have guaranteed income or benefits. In the mind of the salesperson, if they don't close the sale, they don't eat, and their family suffers. (Talk about salespeople being perceived as hungry!) So the salesperson has no other choice than to overcome their discomfort with the technical aspects of that product or service offering and close the sale, which they do. And once this salesperson has closed a technical sale on their own, guess what? They may not feel as compelled to bring in a sales engineer to close their next sale. The technical aspects have been demystified and the departmental silos breached. What is the cost to your company in keeping your sales force in the dark instead of making your technical solutions accessible and understandable?

3. What is the typical scenario that accompanies new projects that come in house? In the "who-is-jumping-through-whose-hoops?" scenario, technical professionals in your organization are jaded about the quality of the contracts negotiated by the sales force. The engineering department anticipates the solution has been overpromised and, subsequently, will be under-delivered at the negotiated price. Or perhaps your sales engineer feels that your customer's design specifications are faulty and should be fixed – after the business is won and the contract comes in house. Either way, design deficiencies are not addressed during the bid or negotiation processes. This status-quo behavior pattern costs your company money once the contract comes in house. If your engineering department makes a habit of second-guessing customer specifications, is this mindset negatively impacting your company's credibility rather than reinforcing technical *value-add*? The customer feels they have

been oversold based on issues that could have been addressed during the sales cycle; your engineers are asking the client to admit that their own technical departments are incompetent.

Rework costs money and profit. In this scenario, your salesperson only functions to close the sale, bring the contract in house, and then run off to close their next sale. Your technical departments have the salesperson and the client jumping through their hoops. Your salespeople are excluded from learning technical information which would allow them to sell larger, more complex solutions collaboratively. Your customers may grow tired of projects that require redesign at increased cost. Your technical departments are justifying their jobs but not reinforcing their value to the customer. You have just provided customers with an easy way out. They may decide to tell your company, "No," or "We'll consider it," or "It's not in this year's budget." Or they may decide to cancel the contract. Ouch!

Getting Unstuck from the Status Quo

The scenarios described in this chapter are caused by a lack of trust in one's colleagues and customers. Lack of trust, poor communication, limited information flow between individuals, and second-guessing decisions are very large elephants that are constantly dragged into all sorts of rooms. This chapter has provided you with a 10,000-foot eagle's-eye view of some of these elephant-ridden scenarios, with particular emphasis on the interface between technical and non-technical departments. You can't move forward until you understand what is holding you back. So take a step back from the situation and keep stepping back and gaining perspective until the real obstacles emerge. You will begin to see the historical and strategic context of all the tactical skirmishes and infighting as they fall into your bigger-picture perspective.

The real issue becomes what you, individually, can do to impact these status quo-entrenched scenarios within your organization. I am not asking you to climb mountains or start a revolution. But sometimes doing just one thing in a non-status-quo manner can cause the people seated around the table to see things differently. Your actions can chip away incrementally at the status-quo mindset in a productive and valuable manner. Slowly the elephants start to disappear as successful collaboration unseats entrenched habits. Your actions invite your colleagues and customers to collaborate, ask questions, and provide insights. Your willingness to lead by example may unlock the aha! potential in everyone involved. Think what your workplace would look like if these small changes started to happen on a daily basis.

REVIEW OF MAIN POINTS

1. A naysayer is someone who systematically obstructs individuals from taking action. Naysayers exist in companies as a means of control. Naysayers may feel they are the voice of reason and caution, and may see their actions as positive.

2. Decision-makers say no when they are brought into the decision-making process too late and with insufficient or irrelevant information to fuel their decision. They say no when they perceive the decision as too risky and disruptive to the status quo that they feel compelled to preserve. And decision-makers may say no simply because they dislike or mistrust the data or individuals from another discipline, whether this is a logical assessment or not.

3. Making your internal and external customers jump through hoops, second-guessing your customer's design specs, applying sales engineers as needed, and causing sales professionals to close sales on a wing and a prayer, are examples of behavior in companies that feel it is more important to reinforce their siloed divisional structure than incorporate collaborative, cross-functional business development processes.

4. Companies whose mission is to preserve their internal status quo negatively impact their ability to be competitive and profitable. Companies that give themselves permission to change incrementally liberate themselves from what is holding them back. They move forward.

SALES-ENGINEERING INTERFACE™ TOOL #2: THE ANATOMY OF *NO*

1. Which departments, if any, are known as the no-you-can't-do-that people in your company? What has happened in the past to fuel this position?

2. Who are the major decision-makers in your organization? What are these individuals' criteria for evaluating a decision? If you do not know the answer to this question, ask internal resources in various departments.

3. Provide examples of each of the following sales-engineering scenarios in your company: (a) apply as needed, (b) thanks, but no thanks, and (c) who is jumping through whose hoops?

4. How can moving one millimeter outside your comfort level impact your role in each of these scenarios? What steps are involved in moving yourself beyond the status quo?

CHAPTER THREE

FINDING THE COMMON DENOMINATORS

It's about Revenue Generation

"Many businesses across post-recession America are asking employees to assume multiple roles, transforming the nature of their work…engineers often must have the financial acumen to figure the profit margins of their jobs and pick materials accordingly…"
~ Paul Davidson, USA Today, July 6, 2011,
"Jacks of All Trades, and Masters of All"

The financial meltdown of 2008 clearly demonstrated, among other things, that no one is essential to their company unless they are meaningfully and legitimately aligned with revenue generation and the business development process. No one has job security, even if they feel their technical expertise and/or seniority are insurance against unemployment. The individual who is ill at ease communicating across disciplines within their organization, or with C-level (corporate) decision-makers at customers' companies, does not bring real value to their company's table. The marketing and sales professional who views their job as an eighteen-month tour of duty until their next promotion provides only short-term, tactical return on investment for their company; they are disposable once company goals are met. The engineer who is not able to make their technical information accessible and understandable to non-technical professionals becomes a liability rather than an asset to their organization.

It's scary out there in the career world. All you have to do is listen to the news to get a sense of the precariousness of the economic outlook for the job market. Companies are working leaner and meaner as they struggle to contain the costs of doing business. They are focused on achieving more with less. Organizations are looking for individuals who have the mindset and capabilities that allow them to be trained to function across disciplines. It's your responsibility to utilize the opportunities at your company to prepare yourself for (a) your next manager, (b) providing enhanced value to retain your current position, or (c) your next job. There's no room for finger-pointing or excuses. *You* have to mean business – and revenue generation – for your company.

"How do you get typically introverted (engineering and operations) minds to communicate with typically extroverted (sales and marketing) minds?...I think technical professionals of the future will evolve and adapt to more of a technical sales personality because of everyone being forced to do more with less...forcing them to keep more than just the details of the design in mind."
~Keith Bradt, Sales Engineering Professionals Group, LinkedIn, 5/16/2011

The common denominator across everyone's job descriptions, whether stated explicitly or not, is revenue generation. Revenue generation drives your company, whether you are in the business-to-business sector (B2B) or the business-to-consumer (B2C) sector. Without a revenue stream, you don't have the luxury of perpetuating the us-versus-them and status-quo mentalities. More important, aligning your perspective and output with your company's business development process at least grows the skill set you offer to your current employer. By enhancing your capabilities to include this broader perspective of revenue generation, you become more valuable to future employers.

Common Denominators Provide Opportunities

"...better ideas emerged when people from different backgrounds came together.... If your group is too homogeneous, it will be less creative."
~ Keith Sawyer, *Group Genius: The Creative Power of Collaboration,* p. 131

Do *you* mean business for your company? Companies need to do a better job of generating revenue by focusing on a fluid and ongoing process that brings in the appropriate people early on in the business development cycle. This process accomplishes the tasks of market identification, product development, and maximizing revenue generation and profitability through optimizing company-wide resources. Conducting business by bringing together technical, marketing, sales, finance, manufacturing, operations, and logistics resources to work together in providing solutions for your customers moves everyone one millimeter outside their comfort level. In the long run, you really have no other option.

Identifying common themes running across all disciplines, and making sure everyone seated around the table understands the terminology and principles of the team, provides opportunities for collaboration and innovation. Deconstructing corporate silos can become a powerful business development tool. Think about how utilizing common-denominator teamwork looks to the customers contributing to your revenue stream. This process allows potential and even current clients to view the breadth and depth of your organization while you solidly demonstrate a collaborative, synergistic corporate culture. Talk about a differentiator from your competition!

This left brain / right brain, technical/non-technical business development process may feel extremely awkward to you and your colleagues. If departments learn to communicate across disciplines, you will collectively realize that siloed infrastructures and status-quo mindsets are unprofitable. Fiefdom-building will be less easily achieved and turf wars less readily tolerated. Think of the consequences of charging a department for stalling a project timeline – similar to what is done on public works construction projects where there is a per-day fine when a project goes past the proposed completion date. If you catch yourself and your colleagues falling back into the us-versus-them mindset, you need only stop the process and reset the communication.

This type of business development-focused model self-selects for non-technical and technical professionals who are receptive to and excel at interacting productively with each other and with clients. You and your internal team members *will* mean business, and should gain access to resources and assets whose skill sets match up with the rigors of the business to be won.

The Typical Status-Quo Business Meeting

"Engineers will always compliment marketing types on how they always have their feet planted firmly in the air. Marketing can't understand why it takes so long to get into production so that new and exciting products can hit the shelves. Neither side exhibits an understanding of the other's issues, development times, lead times. There's an old project axiom that goes 'You can have it fast, good, or cheap. Pick any two.' To me, the solution is to get all parties involved not at inception, but even before that. Having an engineering liaison sit in on marketing development meetings can work really great. And, having Marketers assume a capital role as the project sponsor is a good thing, too."
~ Jim Pfister, Manufacturing Operational Excellence Group,
LinkedIn, 4/28/2011

In most meetings, technical and non-technical attendees perk up and tune in to the conversation only when they hear discipline-specific word cues. We may not be compelled, trained, or required to present findings relevant to anyone else's discipline. We lecture instead of asking provocative questions. We play it safe and survive. Since many of us are engaged in preserving departmental behavioral norms and the status quo, we leave those meetings and return to our offices having heard only what was specific to our discipline. We have our latest to-do list and run back to our cubicles to prepare discipline-specific answers to the siloed-thinking questions asked of us. It's like a dialogue between two television screens facing each other, tuned to different channels, with the volume turned up. What a bunch of noise!

The marketing and sales folks want the technical folks to listen to and understand all of the broad implications of their research. The quantitative data presented by marketing and sales people appears lightweight, abstract, and certainly does not look like it was validated using the stringent formulas technical professionals use to calculate tolerances and eliminate defects. The sales and marketing people perceive the questions asked by the technical professionals as condescending put-downs of their marketing research methodologies. The technical folks need data that is compelling; they are anxious to rush back to the lab and start designing. Engineers want solid conclusions, not extrapolation into a broader population that will be polled in the next round of quantitative survey research!

No one is on the same page during these meetings. Everyone has their own discipline-specific agendas and biases. They communicate this information to attendees using discipline-specific language. Each departmental representative has specific concerns to address and projects that they deem important of everyone's time and attention. Yet no one is really listening to anyone else. And the group seated around the table can hardly be considered a cohesive unit working for the benefit of your customers – you know, those folks creating revenue for your company. How can this stalemate be addressed?

What if the first question asked in the status-quo, cross-functional Monday morning meeting was, "How do we all mean business this week?"

Put Yourself in the Shoes of Others To Establish Your Common Denominators

You devote a lot of time to finding out what's important to your customers. What if you spent the same quality and quantity of time finding out what is important to your technical and non-technical colleagues? Sure, the technical professionals want data, data, data to justify their decision-making. What kind of data are they looking for? If you don't understand the types of data that tip the scales in favor of their making a decision, how can you expect them to make a decision to move forward and devote resources to your project? If your market research data seems too fuzzy for technical professionals, teach them the principles of market research and forecasting so that they understand why your team has confidently reached the conclusions you are presenting. Find a way for your data to talk to their data. In fact, why don't you simply *talk* with your colleagues? Just plain, simple talk; the same type of conversation you would have with your customers, with the same level of patience you would extend to them.

Yes, you and your colleagues work for the same company, but this fact does not ensure that communication and understanding exists between or across individuals and departments. A good assumption to make is that no one is ever really on the same page unless you create that page and arrive at it through hard work, communication, and collaboration.

Try on each other's shoes and walk around in them. You may be more comfortable than you anticipated. You may become motivated by the factors impacting your colleagues' decision-making processes. They may be just as hungry for some cross-functional interchange as you are. Think about how much more well-rounded a perspective you will have when anticipating and preventing roadblocks to decision-making. You may move much farther forward in a project meeting than you did before. The us-versus-them melodrama is slowly eliminated as responsible, honest information exchange is achieved between departments, divisions, and disciplines.

I'm not being idealistic here and proposing a let's-join-hands-and-be-friends approach. However, I am urging you to use your training for more than simply responding to the latest to-do list for your department. Become interested, passionate, and, quite frankly, fascinated by the thought processes you and your colleagues use to (a) interpret information, (b) combine it to assess the issue at hand, (c) determine gaps in knowledge, and (d) find a way to move the project forward. This type of thinking is crucial to moving yourself one millimeter outside of your comfort level.

Adopt a Streamlined Approach to Communication

Thinking once again about your cross-functional or even inter-departmental meetings, how can they be made more productive and participatory? There is a tendency to throw a lot of data at one another. After all, everyone assumes that the more data you present, the more credible your conclusions are. What does this quantity-versus-quality strategy accomplish? For starters, technical professionals are far more comfortable with data than they are with cross-functional conversation. When you throw reams of data at technical professionals, they will start picking it apart! After all, that is what they are trained to do – due diligence and analysis. And when you have a meeting in which everyone is a technical professional engaged in picking apart everyone else's data sets, there's very little communication and collaboration! Everyone ends up playing the my-data-is-better-than-your-data game instead of putting themselves in their colleagues' shoes.

What might result if you confine your discussion of metrics, tolerances, and every other engineering test you can throw against the wall into a one-page summary that correlates the most important variables upon which the project outcome hinges? What might result if you limit your multi-city quantitative market research project to one page and the entire project rests on your ability to communicate your findings in relevant and simple language across multiple disciplines? Would your colleagues find it easier to understand the breadth and depth of your thinking, resulting in consensus on next steps?

Some technical departments are requiring communication via a lean-and-mean format in order to avoid team members slinging discipline-specific language and data at each other. This methodology certainly cuts to the chase! And even though other departments balk at this format initially, they are surprised when it actually makes meetings productive and efficient. Streamlined communication between cross-functional colleagues is achieved by limiting presentations to one page, communicated in simple, easily understood terms. Words are important. And when you are limited to the confines of a single page, you must choose wisely.

> "One page thinking forces engineers to describe our work in plain English, simple English, simple language, pictures, images. This cuts clutter and cleans our thinking so non-technologists can understand what's happening, what's going on, what we're thinking, and shapes us in the direction of customer, of market, of sales. The result is a hybrid of strong technology, strong technical thinking, and strong product, all with a customer focus, a market focus."
> ~ Mike Shipulski, "Shipulski on Design" blog, posted June 29, 2011

When technical and non-technical professionals have to condense their messages to one page, it forces them to be succinct and lean. They must strip down their thought processes to the words, concepts, and principles that are the common denominators recognized across disciplines. They must translate what they mean into terms that are relevant to other disciplines.

One-page thinking prevents us from settling ourselves into the comfort of our own status quo. Consider that even the way we present information to each other is rooted in the status quo of the-way-it's-always-been-done in your organization. Your behavior reflects the way you were taught to present data in school. In the last chapter we discussed moving beyond telling someone they can't do something and telling them instead what they can do. We discussed how status-quo environments reinforce status-quo decisions. With a one-page approach, you communicate in concepts that cross disciplines. You use terminology that is understood by everyone around the table, and you keep things simple.

With the career trend slowly moving towards hiring professionals who can work across disciplines, your expertise may not provide longevity in the workplace if you maintain a siloed, even elitist mindset. Your perspective may cost your company revenue as well if you are perceived as being difficult to work with. Hoarding expertise, know-how, and insights rather than sharing them in a collaborative manner, even when the corporate culture remains traditionally siloed, weakens companies, regardless of their size. Learn to streamline your thinking while acquiring the ability to communicate across disciplines. You may find you are able to enhance collaboration and teamwork, and ever-so-slightly move away from the status quo. You *will* start to mean business for your company!

REVIEW OF MAIN POINTS

1. The common denominator across everyone's job descriptions, whether stated explicitly in their job descriptions or not, is revenue generation. Always be asking yourself whether *you* mean business for your company.

2. With the trend slowly moving towards hiring individuals who can work across disciplines, developing the ability to identify and communicate common denominators between disciplines allows you to focus on moving yourself one millimeter outside of your comfort level. Your colleagues may appreciate your new approach as you *all* start to mean business for your company.

3. When technical and non-technical professionals have to condense their messages to one page, it forces them to be succinct and lean. They must strip down their thought processes to the words, concepts, and principles that are the common denominators recognized across disciplines. They translate what they mean into terms that are relevant to other disciplines.

4. The most valuable salespeople in your organization may be the engineers and other technical professionals who understand the dynamics and language of the business development process, and the sales and marketing people who have developed a comfort level with the technical aspects of your business offering.

SALES-ENGINEERING INTERFACE™ TOOL #3: FINDING YOUR COMMON DENOMINATOR IN A COLLABORATIVE COLLEAGUE

1. Identify a colleague in another discipline whom you would like to shadow for an hour or more. (Technical shadowing non-technical, non-technical shadowing technical, etc.) What makes this individual and their discipline so interesting? Do they have a communication style that you admire or appreciate? From whom of your colleagues across disciplines do you learn the most? Is this an individual with whom you have good interpersonal chemistry?

2. Would your ideal collaborative colleague be able to change places with you in your department for a day? Would you also be in a position to change places with them in their department? Explain the reasons you either are or are not interoperable within each other's departments. Do you understand the responsibilities and decisions that your collaborative colleague makes during the course of a business day? Do they understand yours?

3. Which responsibilities and decisions across your respective disciplines overlap? Are these your common denominators? Which responsibilities and disciplines carry the greatest weight during meetings at which decisions are required? How can you cultivate your ability to incorporate this information and succinctly communicate it in one-page format to facilitate forward momentum during the meeting?

4. Take notes during your time together with other colleagues so you can compare and contrast the responsibilities, output, and throughput required from different departments and disciplines. Create a one-page summary and present your findings to your colleagues.

CHAPTER FOUR
BUSINESS BABBLE AND TECHNO-SPEAK CREATE BARRIERS

Stop Lecturing and Start Listening

"One of the biggest challenges I have found over the years working in and with cross-functional teams is making sure when you are speaking that you are talking in the language the person or people understand. This is also part of the preparation for a good sales person. When you are speaking with a buyer it is one type of conversation. When you are speaking with an engineer it is a different type of conversation. When you are speaking with people on the shop floor, again a different conversation. We need to adopt our conversation style to the specific audience so that communication is clear… The other challenge and which I feel is the most important, but is never spoken about, is our listening skills. We may not understand everything that is said (especially the jargon), but if we can summarize what someone is explaining to us in our own words it makes a world of difference. More importantly it also gains the person's trust and respect which is key to success in today's working world."
~ Paul Hayes, Sales Engineering Professionals Group, LinkedIn, 5/3/2011

Whether technical or non-technical, we treat our disciplines like a competitive team sport – another example of us versus them. Our team sport has a team language as well. We brand ourselves by discipline and by our words. Perhaps we even dress a certain stereotypic way as well. We segment ourselves apart from other competitive disciplines. In other words, we stick to our own kind. This concept of the norm, or discipline-specific status quo, makes for some uncomfortable situations when we interact for prolonged time periods with colleagues from different disciplines.

How many meetings have you sat in where people threw around discipline-specific terminology? It probably starts at those Monday morning gatherings. Have you ever found yourself wondering why you were invited to attend these meetings? Do these types of meetings make you feel uncomfortable or like you are on the outside looking in? Perhaps you interpret the interdisciplinary behavior of the attendees in one of the following ways: (a) they figure I know what they are talking about because I have been invited to attend the meeting, and therefore I must be as intelligent and well-educated as they are, (b) they are treating me as if I am not a member of their exclusive club and therefore not entitled to use or understand their discipline-specific language, or (c) they think I am too dumb to get it anyway, and therefore not worth their time. Talk about some status-quo thinking that creates barriers rather than collaboration!

Rest assured you are not the only person in the room who does not understand the speaker's presentation. Some of the other individuals left in the dark may be the speaker's own peers! The real question to ask is why individuals use discipline-specific terminology in cross-functional meetings. Most of the time, they are reinforcing departmental barriers, the us-versus-them mindset, and, of course, status-quo behavior. The net result is that you and other attendees aren't engaged in listening and participating productively in these meetings. You show up and tune out until you hear a word or phrase that makes sense to you. Most of the time you don't pay much attention until it's your turn to present, which becomes their cue to turn off and tune out.

Does Slinging the Lingo Define You as a Professional?

Are you still an engineer if you don't use techno-speak, or a sales/marketing professional if you don't sling around business babble? Many of us feel it is important and impressive to use professional terminology when communicating with each other. Alternatively, we may be trying to protect our departmental turf from "them," and are instructed or encouraged not to share vital information. We form a habit of using professional language as a barrier rather than as a means of facilitation. We become so comfortable speaking business babble or techno-speak with individuals from our own discipline that we assume everyone knows what we are talking about. What value does professional language carry if no one in the room understands what you are saying except the other engineer or marketer?

If you are talking about the point load distribution of a truss system, you will use language pertinent to that subject. If you are extrapolating results from a qualitative marketing research lead-in project to future quantitative research, you will talk marketing and planning statistics. But are you then translating these concepts for the rest of the folks in the room? Most of us unconsciously overuse words, phrases, and departmental lingo we learned in school and now hear daily in conference rooms, at meetings, and between peers. The unspoken message we receive is that this form of addressing each other is what is acceptable in our organization. This norm is the status quo. Besides, we all want to fit in and run with the pack, right?

Communication is the hallmark of humanity. Presentations are the tools of our respective professions, not a shallow form of entertainment. How many times have you stopped talking, started listening to what you were saying, and looked around the room to determine whether or not people understood what you were trying to communicate? If any of us asked our colleagues whether what we were saying made sense to them, we might be in for a big surprise. People might tell us that we, as speakers, love to hear ourselves talk. They might also tell us that they don't get much out of our presentations.

As business professionals, we develop a habit of relying on business babble or techno-speak not only to differentiate ourselves in meetings, but to protect ourselves if we are challenged and not prepared to address questions from the audience. If we are presenting, then we are supposed to have done our homework. We should have all the answers. So what do we do when we are challenged? We fall back into bad habits. We start using the semantics of our respective professions to cloud things in mystery rather than saying, "That's an interesting question. I will get the answer for you." After all, if you are an engineer, aren't you supposed to be a problem-solver and a know-it-all? And if you are a non-technical business professional, shouldn't you be clever enough to extrapolate an answer that is one level higher than the question being asked? Sounds like entertainment to me – as well as a waste of everyone's time.

What's in a Word?

> *"It's all about point of view. The technical people want everything to be technically correct using the industry jargon. The sales/marketing folks want to tell a story or convey a benefit. The challenge is getting both sides to understand that it's not what they want/think – it's all about getting the target audience's attention and understanding. The techies have to respect what the marketing people are trying to do. At the same time, the story tellers have to understand that even the smallest word choice can completely change the technical meaning."*
> ~ Randy Thompson, Sales Engineering Professionals Group, LinkedIn, 5/2/2011

Business development starts by taking the time to learn and incorporate the language, skill sets, and perspectives of the various disciplines within your company. This results in your making yourself more robust in the workplace and the marketplace. You become a resource for your internal and external customers. This enhanced perspective allows you to appreciate the value of patiently listening rather than rushing to seek tactical solutions and quick fixes to customer complaints or issues arising during the course of a project. Instead of having a corporate culture focused on putting out those daily disruptive fires, you might have one in which everyone has time on their hands for innovation. Do you want to be part of the problematic status quo or part of the improved solution?

> *"A true hybrid, speaks the languages to convey meaning, not literal translation. To do this, their listening skills are at least twice as important. After all, we have two ears and one mouth."*
> ~ Malvern Jones, MBA, Sales Engineering Professionals Group, LinkedIn, 5/5/2011, building on a quote originally attributed to a Greek philosopher associated with the Stoics, c. AD 55-135

Words, semantics, and context of use are important to technical professionals. Perhaps this situation creates the greatest level of frustration when they communicate with individuals outside of their technical discipline. Weigh your words before you use them or write them. Revise or refine the vocabulary you use to communicate with your peers and colleagues across disciplines. Cross-functional communication is a learning process within the context of performing your job. Individuals who see the workplace as a means of personal and professional growth, rather than a place of stagnation due to performing set functions over time, will be at an advantage in this type of business development-focused model. Paying more attention to the words you use to express yourself allows everyone to start meaning more *business* in terms of actions as well as words. Consider how you sound to your colleagues before you engage in techno-speak or business babble. Don't let your devotion to such language hold you back.

Keep in mind that the most valuable business development professionals in your organization may be those technical and non-technical professionals who understand each other's language and discipline.

Use Your Value Proposition To Move beyond the Status Quo

> *"A* value proposition *is a clear statement of the tangible results a customer gets from using your products or services. It is focused on outcomes and stresses the business value of your offering."*
> ~ Jill Konrath, *Selling to BIG Companies*, p. 51

When was the last time you articulated your position by using words and concepts that were understood by the right-brained *and* left-brained folks seated around the table? How you communicate your findings and ideas to your colleagues becomes the fulcrum for either perpetuating the us-versus-them mentality or creating a collaborative culture for your organization. Taking the time to communicate the terminology, perspective, and skill set you bring to the table is not a waste of your time. No one intuitively understands the breadth and depth of what your professional responsibilities entail, including the person in the next cubicle. Why not make the effort to let them in on your best-kept secret?

Salespeople spend a lot of time crafting something they call an "elevator speech" for prospective clients. The status-quo elevator speech is based on what a person would pitch to a prospective customer if they had only as much time as a shared elevator ride in which to do so. The average duration of this speech is about fifteen seconds, and traditionally provides an overview of what a prospective customer might receive from doing business with your company.

I don't have to tell you that most elevator speeches are woefully inadequate because they don't engage the prospective customer! The salesperson finds themselves babbling some well-rehearsed business spiel that is vague and sounds like an advertisement. The prospective customer tunes out, just like you do in your meetings, and is grateful once the elevator doors open! The most important question of all is left unasked by the customer and unanswered by the salesperson: "What's in it for me?"

What would happen during your meetings if you delivered information via a value proposition format? A value proposition has a beginning, middle, and end. The proposition tells how your skill set can help your colleagues achieve their desired output. Your statement quantifies the deliverables your colleagues can expect from working with you and your department. There's nothing vague or inadequate about it. When people listen to you, they want to relate the perceived value of what your skill set represents to what they are trying to achieve. Your ability to simultaneously translate your value proposition to both technical and non-technical audiences can brand you as the real deal.

What if your presentations were structured to support your cross-functional value proposition? Value propositions are not lectures. They are hardly us-versus-them, status-quo conversation-enders. How many of us have heard our colleagues articulate their value to us in one of these ways?:

- "While working on this project with my colleagues, we determined design factors that can profitably support sales and marketing initiatives. We will honor customer priorities by using existing technology and manufacturing capabilities, and deliver a cost-effective solution resulting in a 10 percent savings on production and a 15 percent increase in profitability over the next eighteen months."
- "In collaborating with marketing, we incorporated customer priorities with our technology and manufacturing capabilities, resulting in more effective production and delivery, streamlined order processing, and a double-digit increase in revenue generation."
- "We expedited the sales/engineering collaboration, resulting in a streamlined order-to-cash process due to shortened sales cycles, while delivering innovative design solutions, on time, and at the price proposed."

The ideas exchanged by speaking from the position of a business development-focused value proposition can take a project in a direction no one is anticipating. Differentiating yourself by being an accessible and participatory communicator may be more valuable to your company than hiding in your cubicle behind your professional degree within the company's silo.

Jill Konrath's books, *Selling to BIG Companies* and *SNAP Selling: Speed Up Sales and Win More Business with Today's Frazzled Customers*, are tremendous resources for the thinking behind creating a value proposition. Her books provide templates for developing your value proposition and making yourself a relevant priority to your customers. Although written for sales professionals focused on gaining access to corporate decision-makers, the concept of the value proposition is transferable across disciplines and into the interface between sales and engineering.

Communicating your discipline-specific findings to colleagues in terms that reach across the table makes them listen to you in a different way. These phrases are provocative, and the listener wants to learn more. They want to have you on their team. They want to benefit from the perceived value you bring to your role. Your colleagues will put you on their radar screens. You move yourself, and your colleagues, from a reactive, status-quo mindset towards a more proactive, collaborative perspective.

REVIEW OF MAIN POINTS

1. Limiting yourself to professional language shuts other people out of the conversation. Create a dialogue instead of a barrier. Do more listening and less lecturing.

2. Take inventory of how many different colleagues and disciplines are represented in meetings you attend. Understand their professional languages. Observe the manner in which they ask questions and express themselves to colleagues. Decide whether their styles are similar to or different from your own. Anticipate where there might be communication gaps. Incorporate terms and phrases into your presentations that are understood by everyone.

3. Incorporate your value proposition into how you communicate your deliverables. You are there to create value for your colleagues and your company through the skill set and solutions you provide. Let them know that you are incorporating their perspectives when you provide solutions and recommendations.

4. The ideas exchanged by speaking from the position of your value proposition can take a project in a direction no one is anticipating. Differentiating yourself by being an accessible and participatory communicator may be more valuable to your company than hiding in your cubicle behind your professional degree within their departmental or divisional silo.

SALES-ENGINEERING INTERFACE™ TOOL #4: CREATING YOUR VALUE PROPOSITION BY UNDERSTANDING YOUR CORE CAPABILITIES

1. Do you find yourself speaking in business babble or techno-speak? Do you speak this way all the time or only in certain circumstances? Pay attention to when this type of language becomes comfortable for you. Are you including or excluding others from your conversation?

2. In your conversations and meetings with those from other disciplines, how are the words *you* use to communicate disconnected from *their* semantics? Instead of becoming impatient about not getting your point across, ask your collaborative colleague about the root cause of the disconnect. Perhaps their discipline uses the same word differently than your discipline does. After all, what's in a word?

3. Think about the data you are working on and the meetings in which you will communicate your findings. How were you planning on delivering the data to meeting attendees? Do you feel everyone receiving (reading, listening to) the data will understand what you are trying to tell them?

4. How can you phrase your output so that colleagues outside of your discipline can better understand the meaning and value of your data and how it relates to their own responsibilities?

Part Two

UNDERSTAND YOUR VALUE

CHAPTER FIVE
ADAPT, ADOPT, APPLY

Your Functional Role May Be Different from Your Job Title

Many of us take on professional positions that seem great during our interviews with hiring managers, owners, and C-level interviewers, only to find that it's a different story when we start working. Our day-to-day reality may not align with how the job was idealized during the interview process. *On-boarding*, a term applied to employee orientation or "organizational socialization" (Wikipedia), can involve any activity ranging from shadowing a peer for a few days to an elaborate training process. Regardless of the quality of our on-boarding experience, we may find ourselves immediately placed on teams already in the process of implementing projects. Or perhaps we are sent on appointments with a sales rep who needs a sales engineer to clinch a sale with a difficult customer. Nothing is static; we are expected to fit in with the dynamics of the workplace as quickly and easily as possible.

The easiest way to fit in puts us in the middle of status-quo behaviors and departmental mindsets: "This is the way we do things." While important processes, procedures, and protocols may be involved, the socialization aspects of fitting in tend to reinforce a "do things this way and get along with these folks if you want to stick around in your job" message. As discussed in the first part of this book, this type of perspective invites a lot of corporate and discipline-driven elephants into rooms rather than promoting collaboration. Keep in mind that you may bring your own personal pachyderms and status-quo biases into the fray as well. It can get crowded!

You don't throw out what you learned in school and in former positions when you assume a new one. You carry these assets – as well as baggage – with you to your new place of work. Adapting what you learned in the past to a new work environment, while adopting at least part of your new colleagues' perspective, allows you to apply your capabilities to their performance expectations. Your professional experiences should be additive, rather than repetitive. Are you moving yourself forward or spinning your wheels?

Gaining understanding of yourself as a person of worth involves taking a personal and professional inventory. Before you read one more how-to technical or sales book, or earn one more degree or certification, consider the common denominators you bring to your professional life and your personal life. How you adapt your training to situations outside your comfort level, adopt a cross-functional mindset, and apply this skill set to your professional discipline drives your ability to earn revenue and allows *you* to mean business for your company.

When colleagues or clients ask you, "What do you do?" how do you respond? Do you tell them your job title and assume they understand what you do for your company? Job titles these days are misleading. How many of us speak to a vice president of a company thinking they are the decision-maker, only to find out they are an information-gatherer for the real decision-maker, who has a nebulous title? Your ability to make this distinction early on in the business development process streamlines your successful negotiation of the sale. If you are a technical professional, making sure you understand to whom you are speaking is critical. When you have those peer conversations with the engineering department, you may be speaking directly to the CEO! Your casual comments and willingness to expend your company's billable time and product solutions before a contract has been signed may speak volumes to the individual on the other end of the phone or email. And the message they receive may not be positive.

Early in my career I gave a presentation to a room full of well-titled people such as vice presidents, directors, and general managers. I had a fantastic – and large – PowerPoint presentation, a well-rehearsed spiel, and anticipated wowing everyone. I couldn't miss. But I did. As I delivered my presentation, no one seated around that table was engaged in what I had to say. They were looking at an individual who had never participated in previous discussions and showed up to listen in on my talk. After I finished my "wow" presentation, there were a few polite questions followed by uncomfortable silence, until that individual – the chief engineer – asked a few pointed questions, nodded his head, and said, "Sounds good." Then the table exploded into animated discussion. I sealed the deal later that month.

I was fortunate that everyone trusted the value, energy, and diligence I brought to their company. They were still interested in doing business with me, even though I had just spent the prior two months cultivating relationships with everyone else around that table *but* the actual decision-maker! Why? Because I didn't identify who was the real decision-maker for the solution. I assumed everyone's job titles equated with the decision-making authority they held in the company. Lesson learned. Since then I have never been shy about asking how decisions are made, who makes them, and what information is important to each decision-maker.

It's All about Your Core Capabilities

Business development, as well as contributing to your company's revenue stream, *is* part of your job description, even if you are a technical professional. Just like the individuals seated around that table during my "wow" presentation, getting a new title and perhaps more compensation may be the equivalent of a lateral move. And while you may be fine with this type of arrangement, don't get too comfortable. A lateral move may increase personal revenue, but are you involved in decision-making? Serving in a support function may seem like a risk-free position. However, if you are not a part of the decision-making process, and don't have a direct impact on revenue generation for your organization, your role may be perceived as expendable.

The most important attributes you bring to your profession are your *core capabilities* or, as they often are referred to in industry, *core competencies*. They are your personal as well as professional differentiators. Core capabilities can involve one or more talents, skills, or capabilities that you or your company feels are central to establishing and retaining your position in the competitive marketplace. These are unique capabilities, particularly in how they are delivered to your customers. Service delivery of your core capabilities makes it difficult for competitors to imitate you and allows you to expand your skills and services into new products and markets. Most important, your core capabilities are perceived by end users – your customers – as valuable. Core capabilities, and your delivery of them, make your customers want to continue doing business with you.

Your core capabilities form the fulcrum of your professional development. They are your professional building blocks – the ones you get to keep for life and expand upon. You are not a commodity or a stereotype of your profession. Identifying your core capabilities allows you to understand the unique perspectives and qualities you bring to your professional output. No matter what your function is, where you work, or the type of projects in which you are involved, you bring your distinctive, professional signature to everything you do. Remember those value propositions we generated in the last chapter? Your value proposition communicates how your core capabilities create a valued customer experience.

Our core capabilities – those professional competencies – are not our professional degrees or certifications. They are not your Harvard MBA or your PhD from MIT. While colleagues, your organization, and your clients may be impressed with the perceived rigor of your academic training, keep in mind that they ultimately are interested in what's in it for them. If you are unable to apply and communicate what you have learned in a manner that can be tied to their revenue streams, you may not be making as valuable a contribution as you think you are.

You are responsible for understanding and articulating, in language everyone can appreciate, how your core capabilities provide cross-functional value to your colleagues, your organization, and your clients.

Your Job Functionality Is Earned

A mid-level engineering manager for a large high-tech company contacted me after he was asked to head up a risk assessment committee in spite of his having no prior experience in this area. Meetings were non-productive; there was little harmony or support either for him as a leader or the concept of risk assessment. Management told him his appointment was a step towards promotion. He worked longer hours to achieve results from this committee, yet they resisted his efforts, perceiving a token management agenda.

In our discussions, the manager conceded he was not being compensated for taking on these increased responsibilities. Although management indicated his expanded role made him promotable, the manager equated team leadership with an immediate salary increase. Like many engineers, he loved a challenge and had jumped at the opportunity to take on this additional, high-profile role in his company. The extra hours he was putting into this new role took away from his engineering job productivity. In his enthusiasm to please management, he jumped into what he perceived as the demands of his new role. The biggest risk, it turned out, focused around the team members' biases towards any initiative they felt was being foisted by management onto their already very busy shoulders. Upon reflection, he realized there was little support for this function in his organization. Without a job title and sanctioned responsibilities to reinforce his position, he had no credibility or clout with the other team members.

The manager acknowledged he was role-playing as leader without understanding what leadership meant to the team. He assumed everyone would automatically understand his role, fall in line, and accept his new authority. He conceded that the more frustrated he became with the team, the more upset he was at not being compensated for what had become a real headache. He was in failure mode and there seemed to be nothing he could do about it.

I asked him to consider the value he felt he brought to his company in terms of his core capabilities. What spin did he bring to his engineering function that caused management to assign him his new role? We focused on how this opportunity could allow him to become an expert in this particular facet of the company. He could grow his core capabilities if he would first take the time to identify them. Then he started to understand why management chose him to lead this particular team. He acknowledged, for starters, that he was naturally inquisitive and a good, credible communicator between the various engineering, operations, and manufacturing disciplines represented on the team.

The manager realized he was motivated to educate other team members and enhance their core capabilities. He started by providing them with discipline-appropriate articles and discussion forums related to issues impacting their respective roles on the team. He knew his growing expertise in risk assessment would impact his company's revenue stream and profitability. His approach to the team changed, as did their appreciation of his capabilities. His failure mode quickly turned around into a productive experience in which everyone benefitted. And, in turn, by adapting, adopting, and applying what he was learning, he was *leading* – a skill set he could bring to his next job function, either at that company or a different one.

Define and Use Your Core Capabilities

The engineering middle-manager initially had no clue why he was selected to lead that particular team, but his management did. By taking personal inventory of his core capabilities and expanding them with new roles and functions in his organization, he began to see that he was worth more to his organization than his job title or professional degree. He could impact business development and revenue, which are directly related to risk assessment. His growing self-awareness and cross-functional skill set subsequently made him the go-to person for certain types of functional roles in his organization.

Your core capabilities allow you to adapt, adopt, and apply your perspective and skill set to any role you assume within your organization. So how do you go about determining what they are? The first step is to describe *you* to yourself, without incorporating a job description or specific technical expertise. Not so easy, is it? We usually do not see ourselves as adjectives. We like nouns – engineer, associate brand manager, account executive, systems analyst, etc. Adjectives force us to look at how we deliver on those nouns!

Which of the following descriptions describe *you* to yourself? It's important for you to have a clear perception of your professional self, first and foremost. Create a foundation for yourself by focusing on your core capabilities. Allow yourself to develop a firm sense of how you deliver on your core capabilities. Otherwise, you will be playing ready-fire-aim, trying to be all things to all people as you meet their constantly shifting expectations and criteria for productivity. Do any of these profiles describe you?

- "I like to be left alone to work on the small details. I am a highly discipline-driven, precision-oriented person who likes to complete each task, check it off the list, hand it off to the next person, and head towards another in-coming task." (*Note to yourself:* Am I a stereotypic order-taker who wants to engage in business babble or techno-speak and be left alone to do my work? Am I costing my company money through my lack of interoperability, or am I directly and positively impacting its revenue stream?)
- "I am a discipline-specific individual who always gets things done on time so that they are good enough to go, correct, and complete, but may not represent the ultimate or even the best solution." (*Note to yourself:* Am I an order-taker who is satisfied with being in the middle of the bell curve rather than providing innovative and insightful information?

Is my own mediocrity due to my company's approach to output and throughput, or mine?)
- "I am always second-guessing everyone because I don't think anyone can do things as well as I can. I tend to criticize and re-do everyone else's work." (*Note to yourself:* Am I a naysayer or a control person who wants to maintain the status quo? Does my intense allegiance and loyalty to the parameters of my department or discipline provide value, or cost my company money?)
- "I really get into projects! I like to spin off from the task and extrapolate to see whether I can come up with a really robust solution, even if it is outside of the scope of the original project." *(Note to yourself:* Do I have difficulty staying on task, and therefore waste a lot of everyone else's time and energy pointing me in the right direction? Am I an asset or a liability to a project?)
- "I cross over into other disciplines and pull my inspiration and knowledge base for problem-solving from out-of-the-box, multi-functional concepts." (*Note to yourself:* Do I work in an environment that encourages, recognizes, and compensates me for my cross-functional problem-solving acumen, or am I the exception to their status-quo mindset?)
- "I always work well in conjunction with other colleagues and clients when providing solutions. Projects are delivered on time, as promised, at or under cost." (*Note to yourself:* Am I a team player who benefits from working productively within teams? Do projects run smoothly because I know when to lead and when to follow in order to complement the business development cycle?)

Consider how you apply your knowledge base to what your company needs to grow its bottom line. Are your core capabilities aligned with those of your workplace? Are there opportunities for you to apply your core capabilities and move yourself and your colleagues one millimeter outside of your comfort levels in a non-disruptive manner?

Understanding your core capabilities will allow you to articulate yourself differently to prospective employers, colleagues, and prospective and current clients. Understand that who you *think* you are (usually confined to your degree and your job title) and who you *really* are (a functional and valuable team player, a facilitator, a translator between right- and left-brain ideas, an innovator, an order-taker) may represent two different people! Your core capabilities and how you apply their functionality as an engineer or business professional can provide direction to your career development and impact business development for your company.

REVIEW OF MAIN POINTS

1. Your job title, and those of others, may indicate neither the real role you serve for your organization nor your/their decision-making authority.

2. Your core capabilities are the fulcrum of professional development and career satisfaction. Expand them with every project you undertake.

3. Identifying your core capabilities allows you to understand the unique perspectives and qualities you bring to your professional output. No matter what your function is, where you work, or the type of projects in which you are involved, you bring your distinctive, professional signature to everything you do.

4. Your core capabilities, and how you adapt, adopt, and apply them within the context of your functional role in your organization, allow you to provide value to your company and impact its revenue stream. Business development *is* part of your job description, stated or not, even if you are a technical professional.

SALES-ENGINEERING INTERFACE™ TOOL #5: DETERMINING YOUR VALUE

1. In the "Define and Use Your Core Capabilities" section of this chapter, I created some personalities based on core capabilities statements. Which one or ones define you? Feel free to create your own core capabilities description. Remember to use adjectives!

2. Make a list of your current and past job titles. If you are employed in your first job, include all the other jobs you had in college or as a volunteer. At the end of your list write, "My future job title." For future job titles, Google what you think the job title that you are looking for is called. Ask yourself why this job title is important to you.

3. Create a three-column spreadsheet titled "Where I want to go." The first column should be titled "HAVES." Write the core capabilities you currently have that support your goal of achieving your future job title. The second column should be titled "NEEDS." Write the core capabilities you lack in order to support your goal of attaining your future job title. The third column should be titled "HOWS." How will you bridge the gaps between what you have and what you need to achieve your goal?

4. How does your current job title compare with your future job title? Keep in mind that having a particular job title also involves having specific responsibilities that only you can fulfill! Are you being realistic in setting your personal goals? What is your timeline to achieve your goals?

CHAPTER SIX

ARE YOU AN ORDER-TAKER OR AN INNOVATOR?

Everyone's an Engineer

"John Lienhard, professor of mechanical engineering and history at the University of Houston and host of National Public Radio's Engines of Our Ingenuity, *traces the word engineer to the Latin word ingeniare, which means to devise. Several other words are related to this word, including ingenuity."*
~ Press Release, 2011, "The Root of Ingenuity – The Engineer," National Engineers Week Foundation Press Release, http://www.eweek.org

Whether you have a technical degree or a business degree, you are, in a sense, an engineer. You devise *ingenious* solutions for your company using the tools, methods, and techniques of your discipline. You have to research and document your thinking to validate your positions and conclusions to others. Some of us are far more technical, scientific, and quantitative than others; we use theorems, formulae, and mechanics to extrapolate the significance of data. Others of us like gathering the data, dealing with the subjective and qualitative aspects of the problem, and being hands-on for the "people stuff." Together, we create the experimental protocols, contract the researchers (or conduct the research ourselves), and execute the surveys. We interpret the data or hand it off to the seers who analyze the findings and the implementers who produce the output, so our company can execute the revenue transaction. We *engineer* our professional processes and methods.

Even in companies full of engineers and business majors, everyone certainly is not cut from the same cloth. The traditional corporate culture is full of engineers and quantitative business analysts who won't see the light of day or experience client contact unless there are special circumstances. Some of these highly technical, detail-focused personnel are the day-in and day-out individuals who bring consistency – but perhaps not creativity – to the table. They have an uncanny ability to be the order-takers who deliver exceptionally robust results that inspire creative thinking for the company. These individuals understand their roles. They provide analyses and insight that allow colleagues to deliberate the innovative range of the data's implications.

In evaluating your role in your organization and the value you bring to your company, you must correlate what you *do* with how you *impact* revenue generation. This is true for individuals who are so exquisitely technically oriented that they are recognized for their analytical abilities and are allowed to be their companies' data or information gurus, and it's also true for the uber-salespeople who are legendary for closing major, lucrative deals. There really is no safe place to hide; you cannot exempt yourself from having a role that impacts your organization's bottom line. However, it helps to understand whether you naturally gravitate to the role of order-taker/implementer or to the role of innovator. Either way, you need to develop that appropriate cross-functional mindset to enhance your value to your company.

Understand and Use Your Professional Currency

Your *professional currency* is the worth of your value proposition and your core capabilities to the constituents with whom you work; in other words, what *you* are worth to *them*. I'm not talking about what *they* think you are worth (status quo, biases, old paradigm). I am talking about what *your* understanding is relative to the value you create and deliver in terms of output and throughput (cross-functional mindset, core capabilities, new business development paradigm). In order for you to determine your true professional currency, you have to rid yourself of the status-quo biases involved with your discipline. Just as your job function may be more than your job description, your professional currency may involve more than the traditional expectations of your management. Develop your own understanding of your value. Then become comfortable delivering and communicating it.

> "Emerging is when you use a platform to come into your own. Merging is when you sacrifice who you are to become part of something else. Merging is what the system wants from you. To give up your dreams and your identity is to further the goals of the system. Managers push for employees to merge into their organization. Emerging is what a platform and support and leadership allow you to do. Emerging is what we need from you."
> ~ Seth Godin, http://sethgodin.typepad.com/seths_blog/2011/09/mergingemerging.html

Any sales rep can tell you the status-quo message from their management: after completing a successful sale, they are back to square one. Their slate is wiped clean and they are no better than their last sale. So much for short-term kudos and quarterly sales prizes; they are a commodity in a sales funnel. So much for an elegant process of relationship-building throughout the decision-making process, negotiating the contract, and closing the sale. It's as though management is telling them, "So what? Do it again and we'll raise the bar even higher next time."

Your professional currency as a salesperson is your self-discipline towards continuous improvement of your profession. Do you read books by thought leaders, listen to webinars, ask questions of resources inside and outside of your discipline, engage your customers, document your comfort level with various types of sales methodologies, and understand the sales process – including the front end of that process which involves analyzing the customer's ability to make a decision?

What are your true core capabilities, if in the eyes of others you are only as good as your last sale? Is your process a matter of churning and burning as you inefficiently prospect your way to your next sale via outmoded and ineffective status-quo sales methods? Or do you take the time to understand the qualities you bring to your customers? You just may be a customer-retention specialist rather than a new-business-acquisition rock star. And in this tough economy, companies are finally beginning to realize the value of retaining their current customer bases. What is your value to your organization if your customer retention rate is 100 percent each year, regardless of how many new business sales you close?

If you are someone who excels at sitting in a cubicle and analyzing data or designing a technical solution, do you feel that discrete, technical problem-solving has no direct relationship to revenue generation? Once you hand off your output, is your attitude that the project is now someone else's problem? If you answered yes to these questions, there's a disconnect between your need to play it safe and your company's need for profitability. Understand the processes and methods that generate your incoming data, and the disciplines into which your output is incorporated. Your role in this interface creates your value.

Whether you are a technical or a non-technical professional, take the time to map what you *think* your role is within your organization. Understand the breadth and depth of your professional reach to your colleagues. Your output has the capability of impacting and influencing multiple users across multiple departments and disciplines. How can you make your output more worthwhile to them? When can you make yourself more accessible to them? Would it be effective to develop a more cross-functional mindset that takes into account the different user groups on the receiving end of your output? Order-takers – the really good ones – are critical components of the business development process. Make sure your company understands that *you* understand this distinction.

Let's revisit some scenarios. Do you enjoy leaving the comfort of your cubicle (and your discipline) and translating data for colleagues outside of your discipline so they can use it to create and implement technical or sales solutions? Your serving this type of role involves the risky business of communicating outside of your comfort level and discipline. Perhaps your proposed solutions will lead towards innovation that, upon commercialization, might pay for itself. By more formally and openly incorporating this type of mindset, you enhance your value to your colleagues. Management perceives that you are now using billable rather than non-billable hours, since your output now increases productivity across departments and disciplines. Notice I am not using specific job titles or branding this position as being technical or non-technical. This core capability is not limited to one discipline or another. And it certainly has intrinsic value to your company's revenue stream.

Do you comfortably participate in multi-disciplinary business development discussions with cross-functional colleagues, including business owners inside and outside of your company? Are you intrigued by questions you don't have answers for? Have you identified resources from whom you can obtain those answers? Are you comfortable thinking on your feet and listening rather than lecturing? Perhaps you find yourself combining and recombining existing data with new insight, depending on the context of the decisions to be made. While you are cognizant of the parameters of the discussion, you are not necessarily rule-bound. What is the value of this type of capability to your organization? Are you becoming more directly involved in the primary activities resulting in revenue generation and business development for your company?

Would You Do Business with Yourself?

Be honest with yourself about the type of work you prefer to do. Some of us are order-takers and some of us lend ourselves to innovation. Once we enter the workforce, the practical meets the conceptual, and needs to be articulated to business owners (including yourself if you're the boss!). The business owner will not pass or fail us, or give us first or second honors. They are simply going to choose whether or not to continue doing business with us, as either their employee or vendor. Your value proposition, core capabilities, and ability to articulate them to colleagues, clients, and your employer make you someone people should want to do business with.

In this expanding global economy, simply doing your specific job is not enough for keeping your job or retaining your client base. Corporate on-boarding, sales training, and professional certification programs may be inadequate at best, and more often than not they are disappointing. While academia can prepare us to do a competent job in the workforce, the context and application of our throughput and output in our professional setting is what contributes to a positive or negative revenue outcome. By developing the acumen to identify opportunities in the workplace and the core capabilities needed to address these opportunities, we grow as professionals.

> *While the term "value-add" is often used to describe what is provided to customers, does anyone really know what this term means?*

I heard the following analogy several years ago while attending a sales conference. We were discussing the role of the engineer in most service companies and how difficult it is to engage them in the sales/business development process. My colleague felt that many engineers perceive their functional role in their company as similar to that of a waiter at a restaurant. They hover over the table, waiting for the customer to order, so they can rush off to the kitchen and do the cooking (design). The problem is that the customer may not know that the restaurant (the vendor's company) exists, or be hungry for what the restaurant is serving (the vendor's core capabilities). And the customer won't ever find out how great the food is if the waiter (engineer) only waits around for an order (a salesperson bringing in a contract). The customer wants the waiter to engage them and make them hungry for the menu of offerings (become directly involved in business development). Some analogy!

At this same sales conference, my colleagues observed that most salespeople were no better than talking heads or a brochure with legs (the menu). By the time salespeople got an appointment with the customer – who might be an engineer or another decision-maker – they were "showing up and throwing up" all sorts of information based on their misassumptions that the customer was hungry for their products and services! Like the waiter, they hadn't even taken the time to determine whether there was a contextual need for their offerings (is the customer even hungry for what you offer on your menu?), let alone whether there was any immediate or realistic means of making a decision to buy your solution (order food from your kitchen). Salespeople tended to equate making an appointment ("Someone has time on their hands and I can meet my appointment quota for the month!") with the buying process and business development cycle ("I am sure I can talk them into being hungry for my solution"). What an analogy!

Do either of these scenarios apply to your perceptions of your technical or non-technical colleagues? Everyone is operating inside a vacuum! Technical and non-technical professionals are stereotypically viewed as being order-takers. Thinking of your current company, how feasible and realistic is it to expect professionals to perform isolated functions and hand them off to someone else after they are finished with their tasks? This sales and engineering business model resembles piecework on an assembly line. I don't think you view the net worth of your professional training in this manner.

Would *you* do business with yourself? Incorporate a perspective that allows you to anticipate the needs of not only the person to whom you will be handing off the project (your internal customer), but also the individual from whom you will be receiving output (again, your internal customer). How do you feel you are regarded by your customers – as an order-taker, a partner, and/or an innovator? Input, output, and throughput should be top of mind when you are dealing with your internal customers as well as when you are working directly with prospects and clients. Gradually incorporate the perspectives of individuals up and down your internal organizational food chain. Take the time to determine how your clients make decisions. You will start to build the professional currency that is critical to business development.

Do You Know How To Play Well with the Other Children?

Recently, a colleague of mine in marketing and sales related her frustration to me. Apparently the engineering department was unwilling to relinquish ownership of a project. The company was historically conservative and engineering-intensive, and it felt to her as though the R&D and engineering folks were treating the marketing department like children who could not be trusted with the responsibility of managing their technology. The R&D/engineering folks had come up with a risky idea involving new product development. Now it was time for that hand-off to the department responsible for commercialization and revenue development. The engineering department was reluctant to lose control of their baby – their creation. When discussing the problems with project transition, the technical departments' collective perception was that the project would somehow be degraded once the descriptive, non-technical, "common" language for commercialization and marketing was applied.

The marketing and sales people seated around the table did something remarkable for their corporate culture: they didn't go on the defensive. Instead, they took the time to explain how the technical information would be communicated to the end-user. These non-technical professionals collaborated with their technical colleagues in translating the technical language so that there was nothing lost in translation, but a lot gained in consumer understanding. It was an important innovation for both sides in that conservative company. What started as a status-quo stand-off eventually resulted in collaboration – once everyone kicked their elephants out of the room.

Learning to acknowledge and defer to the expertise of others and let them do their jobs is a real challenge – especially when you feel you know more about the project than they do. Letting go is difficult to do, whether you are on the technical or non-technical side of the table. Whether you or your department have had your way for a long time (us versus them), or you are a newbie wanting to enhance your professional credibility, it's hard to relinquish your role and transition projects. You can't imagine how things will turn out if you do not remain an integral part of the process. You prefer to continue to lead rather than collaborate or follow. You try to inject yourself back into the thick of things without respecting the new process and leaders.

The big question you need to ask yourself when it comes to project handoff is why you want to remain involved. Some mourn the handoff as though they have lost a loved one. Technical folks are concerned that the marketing and sales effort will minimize the features and benefits of their creation. Sales and marketing folks fear they will end up with an over-engineered product that piques the fancy of the R&D folks rather than meeting the needs of the end users – the consumers.

Project hand-off means you take a new seat at the table and serve a different, more collaborative role. This is when you become "all ears" and listen to the language of marketing, sales, business development, and perhaps even an engineering department halfway across the world. This project phase is when you once again become a student. Determine how your involvement in this project has contributed to your core capabilities. You will ask great questions – not second-guessing, micromanaging, or deprecating ones – that allow you to provide even greater value to your team. This is the point at which you gain a 360-degree perspective and return that perspective to your own engineering or sales/marketing department for the next round of new product development. This is the time when you can incorporate what you have learned to become more innovative. Why should your discipline-driven or departmental biases hold you back? Take that 10,000-foot eagle's-eye view of the project and move yourself and your colleagues forward.

It all boils down to using your professional currency, your value proposition, and your core capabilities to engineer trust between your colleagues and comfort with the disciplines and processes involved. Depending on where we sit around the table, we see the same things diversely, differently. So why not respect a project hand-off as an opportunity to learn from your colleagues? They just may know what they are doing after all. I doubt that anyone was ever hired to be a solo act in your company, and business is not conducted in a vacuum.

REVIEW OF MAIN POINTS

1. Whether you have a technical or a business degree, you are, in a sense, an engineer. You devise *ingenious* solutions for your company using the tools, methods, and techniques of your trained discipline.

2. Your professional currency is the worth of your value proposition and your core capabilities to the constituents with whom you work. How are you going about differentiating yourself from stereotypic, commoditized perceptions of your professional discipline?

3. Simply doing your specific job is not going to be enough for you to either keep your job or retain your client base in this expanding global economy. If you were a business owner, would you do business with yourself? If you were a stock, would you invest in yourself?

4. No one was ever hired as a solo act in an organization, even in a business of one person. Understanding when to defer to individuals with complementary expertise, and assume a different or supporting role on the project, is an important element in building your portfolio of core capabilities.

SALES-ENGINEERING INTERFACE™ TOOL #6: DEVELOPING AND USING YOUR PROFESSIONAL CURRENCY

1. Relying only on your value proposition and your list of core capabilities, articulate how you create solutions for your organization. What do you feel is the overall value of your contributions to your company?

2. If you could develop a coin or symbol that represents your professional currency to your organization, colleagues, and clients, what would it look like? Is this a desirable currency? What would you need to do to make this currency more desirable and valuable to your organization?

3. Would you do business with yourself? What are your perceived benefits and drawbacks? How do these benefits and drawbacks correlate with the gaps in your value proposition and your core capabilities?

4. Carry your professional currency in your mind into every touch point you have with your internal colleagues and your external customers. Use your professional currency to add value to these interactions. Keep a record of how you are able to do this.

CHAPTER SEVEN

WHY SOFT SKILLS ARE POWERFUL

What They Didn't Teach You in Engineering or Business School

"[Coworker:] 'Wally, can you review this for any engineering issues?' [Wally:] 'What issues do you think it has?' [Coworker:] 'I don't know. I'm not an engineer.' [Wally:] 'Your request is too vague. You need to tell me what issues I'm looking for.' [Coworker:] 'Did you just ask me to do what I just asked you to do?' [Wally:] 'I don't know. I'm an engineer, not a linguist.'…"
~ Scott Adams's *Dilbert*® cartoon

"The ability to deal with people is as purchasable a commodity as sugar or coffee. And I will pay more for that ability than for any other under the sun."
~ John D. Rockefeller

Many technical professionals feel that undergraduate curricula do not teach them the *soft skills* necessary for sales, marketing, or running a company. I have news for you: neither do most undergraduate business programs. Please stop using this situation as an excuse for having weak soft skills such as writing, presenting, negotiating, listening, networking, mentoring, decision-making, empathy, and patience, among others. These are the skills that many hiring managers feel contribute to *emotional intelligence*, or EI, according to a recent study conducted by CareerBuilder.com (http://www.careerbuilder.com Press Release, August 18, 2011).

Have you ever sat across the table from a CEO during a business development meeting? They throw lots of multi-disciplinary questions at attendees that involve far more insight than providing a discrete technical or marketing answer. Since business development creates the cash flow and revenue stream that fund your paycheck, developing the soft skills required to answer these types of questions might be well worth your time.

In the last chapter, we discussed developing and understanding your professional currency – the value you bring to your job and your profession, your core capabilities, and your interactions with others. In order for your value to be understood and appreciated, it needs to be communicated. And communication of your value and your core capabilities is the fulcrum of your career. If you can't communicate, how can you advance? When placed in this context, those soft skills don't sound very soft at all.

I consider soft skills to be essential components for business development. And business development *is* part of your job function, even if it's not written into your job description. Those alleged soft skills involving communication are as much a part of your professional arsenal as are discipline-specific theorems and methodologies. Whether you are working for a manufacturing company or a service company, you do not practice your discipline in a vacuum, and you don't work in a homogeneous environment. So stop perpetuating the vertical thinking and siloed mindset that are carry-overs from your undergraduate and graduate days or your last job. They are the huge residual elephants in everyone's rooms.

None of Us Took Classes about How To Be People

"Engineers need to have…strong communication skills (listening, speaking, presenting, and writing); knowledge of history of engineering (events, people, successes, failures, etc.); and management and leadership skills (accounting, planning, strategic thinking, ethics, etc.)."
~ David Lourie, P.E., D.GE, ASCE Group, LinkedIn, 2/14/2010

There is too much excuse-making about the shortfalls of education in developing communication skills in graduates. It's not a matter of what you didn't learn, what was omitted from your undergraduate or graduate curricula, or whether you had time to take those extra classes. None of us took those extra classes. But we all took classes involving team-based assignments. What about those discussions and negotiations between team members regarding the relative worth of the data being collected and the analyses being applied? Were you always the worker bee in these team-based assignments, or did you assume a leadership role from time to time? Was your leadership based on your magnificent use of technical or business linguistics, or your ability to coordinate everyone towards a common goal?

Even if we could have registered for these missed-opportunity undergraduate classes, what would the titles have been? "Communication with people from a non-homogeneous sub-population"? Or how about "Coordination of individuals with differing perspectives towards project completion"? These titles sound more like anthropology classes than simple, consensus-seeking discussions with a client or the guy or gal down the dormitory hall from you. Perhaps this perceived communication insufficiency is more a matter of whether or not you are using everything you actually *did* learn in school, because you learned a lot. You certainly didn't learn in a vacuum. You participated in teams and experienced group behavior dynamics. The last time I checked, that pretty much sums up the atmosphere in the workplace.

Those gregarious non-technical colleagues of yours in sales and marketing also are plagued by a lack of communication skills! Would you text a formula to a colleague, Tweet™ a response that could be interpreted in more than one manner, or send an email in shorthand full of those business buzzwords, with a copy list including everyone under the sun? I highly doubt it. And you don't get a free pass out of this discussion simply because you are not using technical language to communicate. As we discussed in Chapter 4, words have meaning, context, and therefore power. How you use words is extremely important, particularly to technical colleagues. What's being lost is the art of using the most appropriate words to convey what you are talking about.

Remember those ACTs, SATs, GMATs, and other entrance exams? Didn't these exams have a semantics and syntax section? It seems that everyone learned enough vocabulary to get a good mark, and then promptly lost the ability to incorporate these words into a sentence after they took these tests. If you don't practice what you learn, you lose it! Using an app like "Word of the Day" might be an interesting means of stimulating your brain. Having a "conversation of the day" with someone who is from that non-homogeneous subgroup at work could be revelatory. Just imagine what you might achieve when your practice conversations become comfortable and natural.

You Will Be Evaluated on Your Communication Skills

"Get out of your chair and go talk to someone in your office rather than phone or email them – the value of face-to-face communication opens the door for deeper understanding, better knowledge transfer and mentoring moments."
~ Warner Coffman, SPHR, Civil Engineering Central Group, LinkedIn, 3/16/2011

The criteria for evaluation on a performance review may seem like a constantly shifting target. Whether you have a solid technical or business foundation, it's critical to develop the skill set for identifying the tactical details of the big picture, and to be able to communicate your findings. If you treat your functional role in your organization as problem-solving in a vacuum, or perceive fulfilling your sales role as piecework or order fulfillment, then you are shortchanging everyone, especially yourself.

You get out of bed every day and function using your senses, and acting and reacting to situations, as you commute, order coffee, buy a newspaper, and do your human stuff to get to your workplace. There are vendors who may not speak your native language, individuals who are hard of hearing, buses that are late, and those just-plain-cranky people who are doing the same thing you are. You get your point across and so do they. So why turn off this ability when you walk into your place of work, marginalizing people instead of creating synergy and collaboration? Perhaps eliminating insufficiencies in communication is more a matter of learning how to communicate with people you don't feel are as smart as you are, no matter on which side of the technical/non-technical continuum you reside.

It's no longer acceptable to graduate and not be able to speak productively to your peers, regardless of their academic disciplines. And learning to speak to others effectively may involve developing friends outside of your academic and professional circles. The dynamics of business are hardly confined to peer conversations. How can you expect to rise through the ranks of your professional organization and garner successful performance reviews and recommendations without effective communication?

You need to regard yourself as a person, first and foremost, and a professional next. There is a lot of talk about authenticity, which boils down to being the same person regardless of the role you assume in a situation. Do you challenge your communication skill set outside of the work place, for example, by mentoring and tutoring grade school students or volunteering for a stewardship activity like Engineers Without Borders® or Habitat for Humanity®? Have you ever placed yourself at the mercy of people with limited technical resources or linguistic barriers, or who have little or no education, but an overriding desire to improve their situation? They'll certainly find a way to communicate with you. No problems there.

When you take away the tools of your trade – your job description, your job title, your academic degrees, and your ego – you have nothing left other than your willingness to communicate. Communication is the greatest common denominator between people. Perhaps the workplace feels like a primitive frontier to many technical and non-technical professionals. What would you do if your annual performance review were in the hands of one of those Engineers Without Borders® recipients residing in a rural village in Peru? Now that's something to think about.

Are Your Customers Comfortable Doing Business with You?

Business development requires more than problem-solving. You need to uncover the behind-the-scenes factors that impact the customer's corporate culture. After all, that is the environment into which your solution will be placed. The best way to practice your powerful soft skills is to listen to what your colleagues and clients are telling you, and respond by asking questions that assure them that you have listened to and understood what they said. And they may want to provide you with valuable information that has nothing to do with your specific solution! Yet your customers may feel that they need to translate their issues into your professional language in order for you to understand what they are talking about. And they might not do a good job of translating because they, like you, are uncomfortable speaking outside of their professional syntax. You – not your customers – need to develop the ability to understand the semantics on both sides of the technical/non-technical table. Make your customers feel comfortable doing business with you.

Asking your clients to talk about what they know best – themselves and their company – unearths a lot of information that can prevent a project from derailing. Customers and prospects are not so much interested in what you can do *for* them as in learning how you can work *with* them as a trusted business partner. After all, you don't live with these individuals and you don't work *for* them. In fact, you only occasionally work *with* them. Your knowledge of what their issues are may be confined to the information other colleagues share with you, what you learn during occasional meetings with them, and what you have read about the company in your research. There are limited opportunities for your customers to become comfortable doing business with you. Make the most of them.

You and your client may not be on the same page even though the conversation sounds positive and agreeable. Your ability to repeat back to your client what you *think* they have just told you, and obtain feedback for validating or clarifying yourself, may be one of the best soft-skill tools you can use to develop cross-functional communication. Asking questions is not a sign of ignorance or lack of professional strength. It's a sign of leadership. By asking questions of your clients and colleagues rather than always providing solutions, you collaborate with them. You acknowledge their expertise in the business development partnership. Your actions make them want to do business with you.

Keep in mind that communication is not information-gathering; it's dialogue. Your speaking patterns need to be natural and comfortable, and they should not make it sound like you are soliciting information to complete a checklist. Communication is understanding the context of decision-making and the real nature of the project objective. Those soft skills are the powerful, essential skills in business development. You cannot ignore their value to your professional currency.

REVIEW OF MAIN POINTS

1. Skills such as writing, presenting, negotiating, listening, networking, mentoring, empathy, and patience, among others, are often referred to as soft skills by technical professionals. These skills form the basis of your professional currency, and are, perhaps, some of the most valuable and powerful skills you can develop.

2. Communication is a collaborative dialogue, not a monologue or a robotic question-and-answer process.

3. If you rely on professional language, business babble, or techno-speak to communicate with your customers, they may not feel you are listening to what they are saying. If they don't feel comfortable communicating with you, they are not going to want to do business with you. How do *you* sound to *them*?

4. You can learn a lot more about your customers' needs by listening than you can by talking. Since you work *with* them, but not *for* them, you will never completely understand their context for decision-making. Listen at every opportunity so they feel comfortable doing business with you.

SALES-ENGINEERING INTERFACE™ TOOL #7: WORDS ARE POWERFUL

1. How would you assess your soft skills? Are you comfortable speaking with colleagues and clients outside of your discipline? Why or why not?

2. Listen to how successful individuals in your organization collect information from colleagues and clients. Do they use a tactical, one question - one answer technique, or is the information obtained through relaxed dialogue and questions that result in free-flowing responses?

3. Now it's your turn. If you have a meeting coming up, practice asking and answering questions that may arise. What will your questions and answers sound like? Develop questions that you plan to ask. Practice some alternate responses that are different from your usual style.

4. Explore some options such as Toastmasters® or a volunteer or stewardship program in your community that will provide opportunities for developing a relaxed and comfortable communication style with individuals who may be way outside of your professional peer circle and comfort level.

CHAPTER EIGHT

WHY DO YOU WORK FOR OTHER PEOPLE

It's Your Script, Not Theirs

Most of us develop a mental image of the drama of a new workplace when we accept a new position. You might imagine a script about fulfilling an important need within the organization. Perhaps you envision taking a key role in a meeting, presenting findings that are instrumental to an important decision, or being a major source of the data used by many departments. When things do not play out as you hoped, it can be the root cause of dissatisfaction with your job, your coworkers, your clients, and your employers. Keep in mind, however, that it's a two-way street. Your company, coworkers, and clients have expectations from you as well.

Perhaps the most important question you can ask yourself is what you want to give to, as well as receive from, your current position. Just as you created your value proposition and list of core capabilities, it is worthwhile to develop a parallel list of deliverables you expect from a prospective employer. Your ability to align your professional needs with what an employer offers you may provide the best insight from which to evaluate a job opportunity, even if it is contracted. Be sure to include on your list the opportunity to learn about and participate in the business development process. Every job you have is an opportunity to learn more about business development.

Regardless of your place in the employment continuum, understanding why you work for other people can be an important part of your career development and the value you bring to your profession. You may be a recently graduated technical or business newbie looking for or experiencing the reality of your first job; you may be a business or technical professional employed under a limited contract; perhaps you enjoy solid tenure with a large manufacturing company, service company, or academic institution. Regardless of your situation, think back to how you landed your first job and whether subsequent positions were repetitions of previous positions, advancements in a direction you were targeting, or serendipitous situations you embraced and flourished in. Could you survive on your own as an entrepreneur, or do your core capabilities necessitate that you work for someone else?

Business Development Needs To Be Part of Your Formula

Rather than falling back into the us-versus-them mentality, try to understand the reasons why the sales and business development functions in an organization are the most desirable, high-powered, dynamic jobs to have. Perhaps one reason is that these positions are directly tied to revenue generation. If you take a closer look at the successful personnel in these key roles, you will find that they either have some sort of technical background, worked for companies known for cross-training their personnel in all aspects of the business, and/or have a business degree that provides insight into understanding the dollars and cents of the organization and how to manage risk. The breadth and depth of their business perspectives differentiate their value.

Perhaps you need to research and target future employment with those companies that offer "best in class" cross-functional development opportunities. At the very least, you can read books about why these companies have corporate cultures that make their employees successful. Research "best in class" companies like the ones discussed in Jim Collins's book *Good to Great: Why Some Companies Make the Leap... and Others Don't*. Take the time to understand how the pedigrees of successful individuals in your own organization, and in those companies for which you might want to work, differentiate them. These successful individuals either combined a technical undergraduate degree with an MBA, or had a finance background that they connected with an engineering degree. In other words, they weren't afraid to cross lines, and they understood the value that hybridized thinking would bring to their value proposition, their core capabilities, and their professional currency.

Successful individuals involved in key decision-making are not nestled in the middle of the pack. They are on the front lines assuming risk and consequences for their decisions. These individuals understand that they neither make decisions in a vacuum nor work in a void, even if they function as independent contractors, inside salespeople, or cubicle-bound technical professionals. They take responsibility for writing their own professional script, and they reject the status-quo mentality that makes others rely on their employers to set them on the path to professional advancement. They control their careers, understand the big picture, and incorporate learning how revenue works within their own and their clients' organizations.

While it's been said that working for a corporation favors individuals who fall in the middle of the performance curve, mediocrity never finds a comfortable hiding place, no matter how big the enterprise may be. Engineers who feel that their post-graduate degrees and certifications ensure job security find they are easily displaced. Business development professionals who are unable to win business find themselves on the outside looking in. With the employment paradigm in flux, your value proposition becomes your responsibility – your script. And your value proposition needs to include an understanding of how your role impacts business development and revenue generation.

Flat-World Choices May Offer Many Opportunities

The contracted technical workforce is always looking towards their next job opportunity. There is no guarantee that the contract will turn into permanent employment, even if you are part of an agency talent pool that places engineers, technical professionals, or sales professionals into organizations that are short-handed. If you are a contracted worker, use this opportunity as a means of fine-tuning the aspects of business and technical expertise that are of interest to you.

The nature and concept of the workforce is changing. Contractual employees, virtual employees, and teams may become the norm for certain industries. Customer service organizations have been outsourcing for quite a while, as discussed by Thomas L. Friedman in his book *The World Is Flat: A Brief History of the Twenty-first Century*. Those who choose to become serial contract employees instead of arriving at that status unintentionally may be creating a new career level prescribed by the current economic climate. For that matter, perhaps your next career move will be to provide outsourced or contracted services to others.

Use contract assignments as building blocks and information-gathering opportunities for determining the industries and professional environments in which you do your best work. Perhaps you have a short attention span, or run through the gamut of your capabilities within a twelve- to eighteen-month time frame. Are you attracted to lofty-sounding job titles to which you feel your professional degree entitles you? Employers will constantly help you evaluate fantasy versus reality! Regardless of what your degree says you should be able to accomplish in the workplace, do you really deliver on expectations – yours and theirs? Your personal attributes may be the reason you have not been more successful at making a permanent footprint along your career path. In other words, make sure you are pursuing your own career path based on your own realistic assessment of your value and capabilities.

Contract employment may feel like something we "fell into" because we needed employment. However, think about the situation as an opportunity rather than an act of desperation. Someone has opened a door for you; it is up to you do decide to enter. In that environment, recognize opportunities and take responsibility for qualifying yourself for advancement. Keep in mind that there is no one recipe for sure-fire success. Avoid thinking about contract employment as making the best of an apparently bad situation; evaluate the terrain, uncover learning opportunities, identify mentors and resources, and network.

We all need to become better at identifying information resources (not just one) and absorbing and using that information. Our capabilities in forming and articulating opinions and insights allow us to better serve ourselves, our clients, and our employers. Your employment should not seem like a continuous uphill battle. The workplace and your profession are where you constantly seek knowledge and incorporate it into your life, both personally and professionally. Keep in mind that business development is part of your job description because you are constantly involved in developing *your* business, *your* profession, for *yourself!*

Do You Know How To Be the Boss?

We have all been in a situation in which we grew frustrated while working for others. We wanted something more, and looked outside of our current organization for the answers. Some of us decide we can do a better job running the business than our employer does. Consider how you might communicate your frustration with your boss when your boss is you or one of your parents!

Before you decide to form your own business, keep in mind that your company has taken the time and trouble to establish a business base from which to draw revenue, saving you the trouble of rustling bushes and beating the pavement to find customers. Your employer has taken the time to develop their professional currency, which can contribute to your own professional reputation. You can state on your resume that you work for so-and-so. Your current employer's name in the marketplace helps you establish your professional credibility.

Working for others reduces your personal and professional overhead. It's a nice way to get your foot in the door as a recent graduate. The employer assumes most of the risk (including liability for your performance).

You will receive benefits and perhaps some bonus compensation tied to corporate profitability. Your current company may even underwrite all or part of your professional certification coursework and higher education, and they may mandate that you continue to work for them for a certain period of time after you obtain your advanced degree. You return their favor and make yourself worthy of their additional investment in you. Clearly there are benefits to working for someone else.

Being your own boss involves generating your own revenue stream, which means understanding how to develop business. Even if you are a displaced professional, becoming your own boss with your own business may be daunting, especially if you do not understand the breadth and depth involved in this endeavor. For those of you in this situation, I encourage you to start learning *everything* required to finance and implement a business before you chart a course as an entrepreneur.

There's a lot on a business owner's plate right from the start. You don't just set yourself up as a figurehead leader and expect things to fall into place. If you are frustrated, learn about the infrastructure and dynamics of your current business and the discipline of running a business instead of getting sidetracked by personalities and water-cooler politics. In fact, take a peek at the SCORE.org website. This site is a fabulous resource of retired experts who are dedicated to mentoring business owners. Having a broader context in which to place your technical expertise, and some great expert resources, can only be a plus if you decide, after all, that you have entrepreneurial tendencies.

Take Your Current Functional Role and Run with It

Business school graduates in entry-level jobs often support the folks who make the real money for their organizations – the account executives. If you are in this situation, understand that your position gives you access to the very data that generates revenue for your organization. You may have opportunities to sit in on meetings between clients and account executives in which you can observe the dynamics of relationships, the types of questions that are important to clients, and how the account executive does or does not facilitate agreement. Start to align this perspective with your professional discipline.

Remember our discussion about job titles and job functionality in Chapter 5? You may find that your job involves being the data collector even though the job description implied bigger things. The actual responsibilities of your employment may not be as important or as big a deal as you thought initially, even though you have a great salary. You may discover that the technical people, engineers, and IT folks have a far greater role to play in policy-setting and decision-making. And if you are like I am, you may find that you are one of the few individuals in your organization who is capable of facilitating discussion, collaboration, and innovation between the folks who always seem to be in two distinct camps – the technical and non-technical decision-makers.

Is your job working out the way you dreamed it would? The common denominator of your professional development remains your ability to align a solid understanding of business development with the core capabilities you bring to your job. Whatever role you decide to play in your organization, including that of entrepreneur, your ability to connect the value of the work you provide to your company's bottom line can become the basis of a very powerful professional value proposition. You *must* mean business.

In the next section of this book, I discuss business planning and how it relates to business development. Before you step back into the status-quo, us-versus-them mindset and tell me that you are not responsible for business development, read on. By now you understand that, in fact, you are very much responsible for business development for your company. So at the very least, you need to understand what a business plan looks like, how to read it, and how to identify areas that might benefit from your expertise. You may be surprised at how many individuals don't understand what a business plan is…including business owners!

REVIEW OF MAIN POINTS

1. Even if you are desperate to find a job, focus on what you want to give to, as well as receive from, your current position. Just as you created your value proposition and list of core capabilities, it is worthwhile to develop a parallel list of core capabilities you expect from a prospective employer.

2. Your dream job should be well-conceptualized and meet your specifications. Your ability to align your professional needs with what an employer offers you may provide the best insight from which to evaluate job opportunities.

3. Ask how you can become more involved in business development for your company. Your interest in driving revenue for your business shows you are willing to partner with management and clients.

4. By understanding and being able to articulate the value that cross-functional thinking brings to your value proposition, core capabilities, and professional currency, you should no longer be hesitant about crossing over the lines of professional disciplines and collaborating with colleagues.

SALES-ENGINEERING INTERFACE™ TOOL #8: ALIGN, ALIGN, ALIGN

1. What is your dream job? Can you visualize what it looks like? Are you working for someone else or are you your own boss?

2. Develop a list of core capabilities that your dream job has to provide for you. What are the functional responsibilities of this job? What are the day-to-day processes of the job? With whom will you interact on a daily basis? Who are your clients? What industrial, business, or consumer segments do they come from?

3. Revisit your value proposition and your list of core capabilities. How well do these align with the core capabilities and specifications of your dream job?

4. How passionate are you about achieving your dream job? How much risk are you willing to assume in order to achieve your dream job? Aligning your professional goals with identified opportunities in the workplace can allow you to give yourself permission to cross those departmental lines.

Part Three

USING YOUR PROFESSIONAL CURRENCY TO DRIVE THEIRS

CHAPTER NINE

EVERYONE IS A CUSTOMER OF EVERYONE ELSE

Who Are Your Customers?

"You can't just ask customers what they want and then try to give that to them. By the time you get it built, they'll want something new."
~ Steve Jobs, Apple

Your customers should include everyone you come in contact with during the course of your workday, including yourself. When you think about it, your interactions with your customers, both internal (coworkers and colleagues) and external (revenue-producing entities that purchase or rent your company's products, goods, or services), represent a continuous business development process. It's your delivery of your vision and your professional expertise, and your successful anticipation of what your customer wants next – rather than now – that create value for your organization. You move from being perceived as an order-taker to serving your customers as an innovator.

In a sense, we are all selling our value to our professional network, and serving them, on a daily basis. Although our coworkers *have* to work with us (that is, they are employed by our employer), they may not *want* to work with us. Ask yourself whether you wish to be sought after as a valuable resource for your company, or to hide and avoid assuming more responsibility than your job description requires. Are you the naysayer everyone in your company dreads working with, or are you a collaborator and innovator? Do you want to be the go-to guy or gal whom everyone wants a chance to work with?

Everyone has customers in the workplace. There are probably individuals at your company with whom you do not want to be associated, but you may need to work with them in order to accomplish project objectives. Then again, there are individuals in your organization with whom you do your best work. How can you create opportunities so that your workday and responsibilities involve engaging more with those individuals and on those projects for which you produce your best output? And how can you gradually shift the balance of your workload so that you increase your value rather than remaining static within the status quo?

You Are the CEO of Your Core Values

Do you think like a CEO? Perhaps you should. In a sense, you run your own business – your current job and, in the long run, your career. When it comes down to it, you are in charge of your life. Tom Peters first wrote about this idea in the late 90s, with his "The Brand You" mantra. Dan Schawbel took this concept into the digital millennium with his focus on personal branding – the subject of his book *Me 2.0*.

I respectfully add a slightly different spin to these two concepts: You are your own business. It is your business goal to run *your* business very well. The business of *you* is based on those core values we worked on establishing in Part Two of this book. Your core values are the fundamental basis of who you are as a person and as a professional. You brand yourself and you establish your personal brand based on those core values. How you serve your customers is not only an extension of your professional expertise, but of your core values as well. Your business, in more ways than one, is *you*. It's not someone else's responsibility to run your business – it's yours.

Regardless of where you sit in your company, business development *is* part of your job description. You are utilizing your core values to develop business for your personal brand regardless of whether or not you feel you are in a secondary, unimportant role. Your perception of yourself needs to be consistently delivered to your colleagues and customers, or else there is a disconnect. That is why it is so important for you to have a firm handle on the fundamental principles, values, ethics, and perceptions that make up the core values you bring to your workplace and to your life each day.

You Are the CEO of You and Your Career

Everything you do professionally focuses on your customers. They are the entities fueling cash flow for your company. You need to be aware of the impact of your actions on the customer acquisition and retention process, and aware of who else in your organization is involved in this process. No one is absolved of their responsibility to this cycle, even technical professionals!

When you attend meetings, you are working with individuals who, like you, are the CEOs of their own careers, whether they realize it or not. What types of conversations do CEOs have? For starters, their perspectives of situations and deliverables are focused on their companies' revenue streams. Billable and non-billable hours are one of many crucial elements that impact profitability. Without cash flow there is no company, and you are out of a job. Business development is a dynamic playing field.

In his book *Selling to VITO™, the Very Important Top Officer*, Anthony Parinello characterizes VITO™ as the CEO, president, or owner who has many dynamic characteristics, including "vision, understanding of the 'big picture,' decisiveness, creativity, openness to new ideas and a willingness to take calculated risks." (p. 46.) If you are the CEO of your job function, how proud are you of (a) your job, (b) how well you perform your job, and (c) the company for which you work?

Bring that CEO-level perception into meetings and conversations with your peers, colleagues, and clients. Shift your focus from the specifics of your job responsibilities and professional expertise to the deliverables of an individual who is responsible for the bottom line. You are your job's CEO, and your performance is grounded in your core values.

Learn about the Big Picture: The Business Plan

Since you own your job, what does this VITO™ big picture look like? What are the big-picture issues that impact not only your job, but the jobs of your coworkers and the goals of your company? If you aspire to moving up the food chain at your organization, do you have a comprehensive understanding of not only what the responsibilities of a new position entail, but also the footprint that position has within your organization?

The document that gives you the best perspective of the big picture is your organization's business plan. This single document connects all the dots. The reader (including you) is provided a comprehensive overview of how all the major functions of the company are supposed to coordinate to generate and perpetuate its value and revenue stream.

Whether you are a technical or a non-technical professional, there's no excuse for not understanding what a business plan is, the data required to write it, and how to interpret this information for planning purposes. You might find yourself running your department or division, starting your own company, or inheriting your father's or mother's small business. The first document a bank will ask you for – as well as prospective big-time customers – is your business plan. It's the first document I ask my clients to provide when I commence working with them.

Learn everything you can about how your business is run, who runs it, how decisions are made, where traditional sources of revenue come from, what happens if there is a shortfall, and what your opportunities are for playing a role in this big picture. Yes, you may be a cog in a wheel, but who told you that you couldn't understand how your cog can be more functional within that wheel? You may end up educating yourself into your next job, let alone advancing your career. At the very least, it makes your job more interesting and, in turn, makes you more interesting to your colleagues – your internal customers. If they benefit from their relationships with you and the information and perspective you provide, they will seek your assistance again and again; they will become retained or repeat customers. Isn't this what business development is all about?

Share Your Information, Because Nobody Can Deliver It Like You Can

While your newly acquired information may cause colleagues to sit up and take notice, share your information resources so they can also access data that may enhance their value. Creating an "I know more than you do and I'm not telling you" atmosphere creates strife. It will take you, your colleagues, and your organization right back into us-versus-them mode. You want to focus on creating value. Your target is collaboration, not competition.

When you think about it, you are creating distinctive deliverables based on your individual professional currency and core values. Keep in mind that nobody can deliver on your job like you do.

The big-picture perspective you incorporate into your throughput assists your colleagues and your company in their output. If you want to enjoy an enhanced workplace experience, make the product of your job delivery something that is sought after. You just may morph into the go-to guy or gal for your organization. That's a concept that your internal customers can easily buy into.

When you share information, you collaborate. When you collaborate, you share a dialogue with your internal customers. That style of interaction may represent a shift in the workplace status quo. That simple act of incorporating information which gives you a big-picture perspective may jolt everyone out of their own complacency, including your CEO! Never underestimate the reach of your big-picture delivery on your job functionality and the perceived value you provide to your colleagues and your organization.

REVIEW OF MAIN POINTS

1. Incorporate a customer-centric approach to the workplace. Your business day should be focused on acquiring, serving, and retaining internal and external customers.

2. You are all business development specialists. In a sense you are selling your value to your professional network on a daily basis. The goal should be one of collaboration rather than perpetuating a siloed mindset.

3. You are the CEO of yourself. You run the business of yourself based on your core values. By thinking like a CEO, you take control of and responsibility for your own business and deliverables. Your deliverables become your professional currency and create value for your organization and your clients.

4. Nobody can deliver on your job like you do. The goal is to create value based on the distinctive deliverables of your professional currency. Put your unique spin on what you bring to your job functionality and the deliverables that internal and external clients can expect from working with you.

SALES-ENGINEERING INTERFACE™ TOOL #9: YOU ARE YOUR OWN CEO

As you read over Chapters 10, 11, and 12 about the elements of a business plan, put yourself in the shoes of a CEO (of you and/or your company). Start by asking yourself the following:

1. Define who your internal and external customers are.

2. What opportunities exist for you to sell or articulate your core values to your customers? Do you take advantage of these opportunities, or do you feel uncomfortable doing so? Why?

3. You are the CEO of your career. You are the pilot of your own plane. What does your career look like? Are you working for yourself or for someone else?

4. How well do you share information with others in your company? How well do you collaborate with your clients? What are your strengths and weaknesses in this regard? Do you know how to be the boss?

CHAPTER TEN

BUSINESS PLAN 101

Not Everyone Has a Business Plan, But They Need One

"Those who build and perpetuate mediocrity…are motivated more by the fear of being left behind."
~ Jim Collins, *Good to Great: Why Some Companies Make the Leap… and Others Don't*

The first question I ask my clients is how they would describe the relationship between their sales and engineering departments. The second question I ask them is whether they can provide me with their current business plan. You'd be surprised at how many established companies do not have an updated business plan, either partial or complete; and these companies range from entrepreneurs and start-ups, to established small and mid-sized organizations, to large corporations. The last time they created a business plan was when they were just starting out, looking for markets and customers, and wrote a business plan in order to obtain financing. Once their businesses were up and running, these documents were filed away and forgotten. But the business development environment isn't static, and neither are business plans. They are not set in stone. Things change. Ask anyone who survived the 2008 economic meltdown. I'd say there's a problem here.

Why don't these companies have updated business plans? I was so surprised the first time I asked for this important document and started to receive piecemeal information and data from other types of plans that had nothing to do with a business plan. A business plan is one of the most important documents you can create for your company. As a technical or non-technical professional, you are the shepherd of your LinkedIn profile, your resume, your core values, and your online personal brand. After all, this documentation is the vehicle for your professional currency. A business plan does the same thing for your company. As the CEO of you, your core capabilities, and how you apply them to your career, you need to understand the type of information the CEO of your company is responsible for providing to investors, board members, customers, and all other parties interested in doing business with your company.

Why is a completed and updated business plan missing or incomplete for many companies? The data these companies perceive as forming the basis of their business plans is either (a) not readily available, (b) comprised of data generated from the payroll software, or (c) incomplete because the business plan took too much work. If you run your business with a day-to-day mindset thinking it's a glorified version of running your household and paying your monthly bills, you need to recalibrate.

Your business may have been started in your garage or at your kitchen table, but there was a critical point (usually a function of volume of business and cash flow) at which it became a bona fide business rather than another household you were running. Most people get caught in a bind at this point; the status-quo habits and perspectives that always worked in the past are no longer sufficient for running their businesses. Such businesses – often sole proprietorships or established family businesses – have scalability issues. Everyone is still running the business from the kitchen table, but they are now competing with the big boys and girls. The time for them to think like CEOs is long overdue.

Do you work for a sole proprietorship that has grown into a small company? Do you work for a family business that has grown into a mid-sized corporation with many family members still at the helm? Regardless of the size of the company for which you work, understanding its big picture allows you to incorporate management's perspective into your job function. And you are making management's job easier by incorporating the issues concerning them – as business owners – into the output from your job. By understanding the elements of a business plan, you are expanding the value you provide to your company. So let's get started on what a business plan is and is not.

Why a Business Plan Is Not a Strategic Plan

Many companies provide me with their strategic plan when I ask them for their business plan. There is a big difference between the two. A strategic plan, generated by a team, is typically related to the strategy and milestones that the individual units or divisions will seek to achieve over the next fiscal year (or more). A strategic plan is not a diagnostic tool for the entire company. It is a forecasting tool for anticipated accomplishments and deliverables by divisions and departments on an annual or longer basis. The objectives may be more speculative than they are realistic.

A business plan, which can run up to twenty-five pages in length without appendices, is a company-wide diagnostic tool that describes the fundamental strengths and weaknesses of your company. This document connects revenue stream to the marketing, sales, and competitive landscapes in which your company functions. In addition, a business plan assesses the operational, personnel, and cash-flow aspects of why your business is in the shape it is in. It's the annual check-up for your business to make sure it is healthy enough to participate in business for the following year.

There is no wishful thinking involved in a business plan. If you feel you are poised for growth, constantly refer back to your business plan to see whether this strategy is reasonable for your company. The data that is part of this document allows you to answer this question before you engage in a lot of planning meetings around a great idea that may or may not have a sound business base. If you want to grow, you will turn to your business plan. Determine whether you are dealing in fantasy or reality, and whether cash flow, sales, manufacturing capacity, and your overall financial debt service indicate an opportunity for growth.

If you are an entrepreneur, you must ask yourself whether you want to license your technology to someone else or whether you want to be involved in the entire process and possibly become a business owner. And while I will be discussing the business plan as it relates to established businesses, it makes sense for entrepreneurs to be thinking beyond obtaining initial financing for proof-of-concept to a strategy that addresses what happens next after your technology or product gains traction and interest from the investor community.

Let's say you want to manufacture a product using a new technology you've developed or offer a certain type of custom fabrication service. Your business plan allows you to access information that will determine whether you have the cash or credit rating available to make necessary capital purchases, hire new personnel, or create a limited partnership with a company that complements your expertise. A business plan tells your customers and creditors whether or not you are positioned to make a new product or service offering.

Based on the previous chapters, you are now aware of the breadth and depth of the factors that impacted your past personal and professional successes and those that are impediments to your future growth. Going through the process of creating or updating a business plan reinforces this awareness. You can assess the gaps in your professional core capabilities to determine whether or not you are able to move on to the next level of career advancement. A business plan provides the same type of insight for businesses.

Why a Business Plan Is Not a Matter of Inputting Facts and Figures into a Templated Document

You can Google the words *business plan* and come up with a number of templates. The SCORE website at www.SCORE.org and the U.S. Small Business Administration website at www.sba.gov have a number of great business-related templates and educational materials that I recommend (including business plan templates for entrepreneurs/start-ups and ones for established businesses). However, don't think that creating a business plan and understanding how to use one is simply a matter of plugging numbers into spreadsheets and templates. I can't tell you how many entrepreneurs I coach who feel that writing a business plan is a matter of collecting data. This attitude is even reflected at business competitions I have judged!

You must understand the background for the numbers generated and information you are providing in such an important document. That's probably the reason why so many companies have given up on this exercise – there was too much information called for and too much care needed to oversee the project. Creating a business plan takes time – at least several weeks. It is similar to a thesis for your business. Don't shortchange yourself. Many companies opt to work with a coach or consultant in developing and preparing a business plan. This move is prudent, since the coach will ask provocative questions, get you to see your company from a number of different perspectives, and assist with – if not create – the business plan in its entirety.

Once completed, you and your company are responsible for your business plan's implementation, which includes getting employee buy-in and pitching it to financial institutions and potential investors, if relevant. It's important that you understand the elements of this document and how to interpret the details. Your company is the physical embodiment of its business plan, just as you are the physical embodiment of your resume. It is your responsibility to be actively engaged in generating this document; it's not just the consultant's job.

Standard Elements of a Business Plan

Whether you are working with a consultant in generating a business plan or have gone online to obtain a generic template, there are basic elements you need to include. You may update your existing business plan, if you have one, to include current trends in your industry or the service you provide. However, at its core, your business plan must include and address these basic elements:

1. Executive Summary
2. Company Overview
3. Status of Work
4. Industry Analysis
5. Customer Analysis
6. Competitive Analysis
7. Marketing and Sales Plan
8. Operations Plan
9. Business Model
10. Milestones
11. Revenue Model

When providing either the entire document or a portion of the plan, make sure that the document you provide is in PDF format so no one can alter any data that you have included in the plan. Retain the master document in Microsoft Word format so you can revisit and update it annually.

Section 1: Executive Summary

Before you dive right in, here's a caveat: This section should be written *last*, after you have written your entire business plan. The reason this section is written last is that by the time you have collected, analyzed, and absorbed the data for your business plan, you can easily summarize it. Until then, the task seems daunting. So spare yourself a lot of anxiety and save this section for last.

The Executive Summary includes an overview of what the reader can anticipate in the complete document: (a) your business overview, (b) success factors, and (c) your financial plan.

When I review business plans as a judge or reviewer for grant applications and business competitions, I turn to two sections first: the Executive Summary and the Revenue Model (which includes the financial statement). These subjects are usually addressed in the first and last sections, respectively. And these critical sections tell me whether a company really can do what it says it can do. A business plan builds professional credibility. Financial institutions and funding sources such as venture capital firms want to make sure that you can "walk the talk." They use your business plan to validate your company. This is not a casual exercise, so save writing this important Executive Summary section for last.

I also recommend that you start concurrently gathering data for the Revenue Model section of the business plan as soon as you start generating material for the other sections. While these data appear at the end of the document, it takes time to collect this information. And the business planning process causes you to ask many questions that have a financial basis.

Business Overview

The Business Overview section of the Executive Summary provides the reader with a picture of the current state of your business. It is a description of who you are, how long you have been in business, in which business segment(s) you operate, and additional elements or capabilities you offer to your customers. Your Business Overview summarizes the framework, or base, upon which your company operates.

The Business Overview can briefly review (a) products, technology, capabilities, and/or services *currently* offered by your company, (b) intellectual property, (c) technology risks, and (d) product development plans. You will expand the details of your Business Overview within the main body of the business plan.

Success Factors

The Success Factors section briefly describes why your company is qualified to succeed in the marketplace. Success factors are not based on fantasy, but on the reality of what your business is based and built upon. Be specific. Don't use boilerplate verbiage such as "providing price, quality, and value" to your customers. Take stock of your specific successes to date and determine the common denominators across these successes. Consider the differentiators that have facilitated your company's success and longevity in the marketplace, stability, and perhaps growth during economic meltdowns. Ask customers why they work with you and take note of the key factors they tell you, including capabilities, human assets, ethics, and commitment to continuous improvement or employee training. Does this start to sound like your core capabilities and your core values?

Financial Plan

The Financial Plan section of the Executive Summary is a brief summary description of what your pipeline and revenue stream will look like over the next three to five years. It should justify why someone would want to fund you or do business with you. If you are actively running a thriving business (which you have briefly described in the Success Factors section), what are the common denominators that contribute to stability and profitability across various line items? If you are in a volatile marketplace, again, you can determine which elements of your company directly and indirectly impact the financial health of your business.

Section 2: Company Overview

The Company Overview is a one-page description that can serve as a stand-alone document you can provide to a purchasing agent to demonstrate qualification of technical capabilities, at a funding meeting with an investor, as marketing collateral, etc. It includes, but is not limited to the data in the following list, and serves as a snapshot of the information that will be developed and discussed in the main body of the business plan.

1. Date of formation

2. Legal structure (LLC, S-corporation, etc.)

3. Office location (headquarters, divisions, etc.)

4. Business stage (start-up, mature business serving customers)

5. Services launched (capabilities, products, and services currently being provided)

6. Revenue milestones

7. Key partnerships (your partners or owners, and years of partnership if not current partners/owners)

8. Key customer contracts (major customers contributing to revenue)

9. Key employees (management team)

10. Total number of employees

11. Commercial and Government Entity (CAGE) code if appropriate

12. Data Universal Numbering System (DUNS) number if appropriate

13. Employer Identification Number (EIN)

14. International Organization for Standardization (ISO) and other relevant certifications

15. North American Industry Classification System (NAICS) code(s)

16. Professional memberships

Section 3: Status of Work

The Status of Work section is a brief, succinct, and confident overview of the way things currently are. State the status of your work honestly and accurately. If you have contracts in house, summarize their impact in terms of how long they will tie up in-house production capacity and provide revenue to fuel your company's cash flow. If new customers or new projects represent a percent increase in either numbers of customers or revenue from previous years, briefly summarize the change in this section. If your pipeline has a multi-year backlog, state the projected profitability for each year based on the reality that these projects are actually in house rather than out for bid.

REVIEW OF MAIN POINTS

1. A business plan is a company-wide diagnostic tool that describes the strengths and weaknesses of your company and connects your revenue stream to the marketing, sales, and competitive landscapes. A business plan also assesses the operational, personnel, and cash-flow aspects of why your business is in the shape it is in. It's the annual check-up for your business to make sure it is healthy enough to participate in business for the following year.

2. The Executive Summary section of a business plan is written last, after all data has been collected and all sections have been written. Typically this section, along with the Revenue Model and financial statement, is read first by reviewers. The Executive Summary and Revenue Model sections are the most important of your business plan and must be written carefully.

3. The Company Overview section is a one-page description that can serve as a stand-alone document you can provide to a purchasing agent, as part of your marketing collateral, for an initial funding meeting with a bank, as qualification of technical capabilities, etc.

4. The Status of Work section is a brief, succinct, and confident overview of the nature of your business and the work filling your current pipeline. It is based on the reality of the types of revenue generated by your company in the past and how repeat and new business from customers is propelling your company forward.

SALES-ENGINEERING INTERFACE™ TOOL #10: THINKING ABOUT YOUR BUSINESS PLAN

1. How would you prepare an Executive Summary for your company? In which departments would you find the information you need to prepare this important section?

2. There's a lot of confusion as to which individuals should form the team involved in the business planning process. Even the experts disagree! If you were to pick a team of the most knowledgeable and experienced individuals in your organization, who would they be, and why?

3. What factors do you feel are responsible for the success of your company? How are these factors reflected in the current status of work for your company?

4. How comfortable do you feel about gathering and understanding your financial data? Which colleagues can best help you become more comfortable with this information?

CHAPTER ELEVEN

YOUR INDUSTRY, YOUR COMPETITORS, AND YOUR MARKETS

"Already, companies that speak in the language of the pitch, the dog-and-pony show, are no longer speaking to anyone."
~ *The Cluetrain Manifesto,* by Rick Levine, Christopher Locke, Doc Searls, and David Weinberger

Section 4: Industry Analysis

The Industry Analysis section of the business plan consists of at least a Market Overview section and a Relevant Market Size section. The Industry Analysis provides the rationale for why you want to start, or started, a business in a particular marketplace, and why you continue to do business in that sector. This section of your business plan must describe your industry, reinforce your understanding of the dynamics of this industry, and communicate your insights for developing business in this industry. If you need resources to gather this type of information, you can purchase online reports, which are very expensive, or you can take advantage of the wealth of information that is available for free online. White papers from relevant industry journals and published articles from industry opinion leaders are easily accessible. Sam Richter's website and book *Take the Cold Out of Cold Calling* (both at www.samrichter.com) are great resources for gathering relevant industry information.

Market Overview

The Market Overview section of the Industry Analysis allows you to determine whether there is a reasonable market for your products, services, and capabilities. No matter how excited you are about your entrepreneurial business concept, the reality of doing business rests on whether there is a market to support your idea and how difficult it is to navigate this marketplace. Mature companies often have so many customers they are serving that they have never taken their own pulse to determine whether or not it makes sense to continue doing business in their current marketplace. I often ask them where they would be today if all of their current customers went elsewhere.

Relevant Market Size

The Relevant Market Size section analyzes how much of this marketplace and market share you currently own and how much more you can realistically acquire. "The sky's the limit" is not a phrase any reviewer of a business plan wants to read in this section of the document. If you are in a niche market, acknowledge this. If you are a generalist, and therefore at the whim of a price-conscious client base, expansion of your position in the market may be a transient goal, at best.

The completed Industry Analysis allows you to realistically assess where your place is within the market based on your capabilities and deliverables. This section provides your company an opportunity to gain insight about whether your goals are lofty or feasible. You will discuss the market opportunities, and therefore the drivers providing the impetus for your company to develop and retain business. Be realistic about what you can achieve.

Section 5: Customer Analysis

The Customer Analysis assesses target customers and customer needs. Business development does not just happen. You need to understand how you have attracted a particular type of customer in the past and whether or not it makes sense for your company to continue doing business with this type of customer in the future.

Every aspect of your business needs to be customer-focused. The Customer Analysis portion of your business plan is an important means of reinforcing why you do business with your current customer base. Often this section of the business plan is revelatory, since it aligns finances and revenue stream with marketing, sales, and business development. Are you serving the right customer base?

Target Customers

Your target customers are the key to the success of your business. Are you aiming in the right direction, or do you have a ready-fire-aim business development approach? Asking your current and former customers why they do or did business with you will let you know whether or not your business plan is aligned with pleasing these customers, regardless of whether this strategy is profitable or not. Perhaps the customer segment you currently serve does not reflect the industry segment you are trying to capture and possibly own. Determine what attracts your current customer base to your company. Be prepared for answers that may not be what you want to hear. Your business may or may not be headed in the best direction.

While it seems daunting to approach former clients for feedback, these are usually the most insightful conversations producing the greatest input. Whether you are technical or non-technical in discipline, this type of information can signal an opportunity or a red flag, which you can pass on to your internal teams. Your former clients may apologize for not being able to continue their business relationship with you due to the economy or behind-the-scenes factors impacting their decision-making. Perhaps their company was acquired by another entity during the process of awarding your company a project, and everything was put on hold. They may appreciate your reaching out to them for their honest opinions. Conditions may have changed since they initially ceased doing business with you; there may be an opportunity to re-engage them as clients once again. You don't know until you ask. Ask.

Customer Needs

Your conversations with past, current, and future customer segments create a Voice of the Customer hierarchy of needs that you can use to assess which types of customers best fit the capabilities, products, and services your company offers. If you are thinking about expanding into new marketplaces, use this list of customer needs as a roadmap for determining cost of entry and whether new marketplace development is worth it in the long run in terms of manpower, timelines, and quality.

Compare the defined needs of your current top ten customers with those anticipated from your top ten targeted customers (prospects). Look at the actual and potential total gross revenue generated and the number of years doing business with your company. Compare this information with the number and nature of projects completed per year, risk factors associated with this client, and their general financial status. Do they merit continued favored status with your company? Consider the hassle factor involved in working with these clients; are they or will they be a pleasure to work with collaboratively, or will they make you and your employees miserable?

Section 6: Competitive Analysis

The Competitive Analysis section assesses, at minimum, the following five factors: (a) top competitors in the industry, (b) top competitors in a niche industry (minority or woman-owned, veteran, HUB-zone, etc.), (c) direct competitors, (d) indirect competitors, and (e) competitive advantages of your company. Whom you identify to be your competition, and the arena in which you want to compete, may not be aligned with your current Status of Work. The sooner you align fantasy with reality, the better.

The Competitive Analysis, first and foremost, looks at your company's strengths and weaknesses. It helps you decide how to capitalize on your strengths within a realistic market space of your definition. Understanding what your company does very well and on a consistent basis provides you with your competitive arsenal. This understanding also allows you to evaluate how to present yourself and your capabilities to new and existing clients. This information is your competitive advantage.

Your Competitors

If you ask your boss (who just may be yourself) who your competition is, they may provide a list of three, four, or more top competitors – companies who bid on the same projects your company bids on. Just because these companies always seem to compete with your company for the same business does not mean they are the actual top ten competitors for your market niche. You may be thinking too small or too locally. Since you've already completed your Industry Analysis and analyzed customer needs in viable markets, you may find out that you are pointing your resources towards the wrong target customers and actually competing against an entirely different set of competitors! Talk about ready-fire-aim!

Some civil engineering firms landed in this predicament due to the financial meltdown of 2008. They found themselves having to diversify and compete against companies whose revenues primarily came from bidding on public works jobs. The more customized engineering firms were not used to the fee structure of this type of bid process or the price-dominated basis of awarding contracts. While it is commendable and necessary to work to keep revenue coming into your company, if your entire operation is based on the habits, practices, values, and skill sets of serving a different type of client set with different criteria for awarding contracts, you may be in for a big surprise. Stay true to your industry and marketplace, providing you have analyzed these areas. Ask yourself whether it makes sense for you to continue working with your existing target markets, or to diversify into new areas.

Direct and Indirect Competitors

While you may know the names of the big and little guys and gals who dominate your industry, they may not be your direct competitors – the companies which are your true competition. Sometimes you are so focused on competing in market spaces dominated by the big guys, or bidding on sexy, high-profile projects, that you chart a course in a direction that is not suitable to your company's core capabilities. Some of your identified competitors are simply too big – or too small – to handle projects of a certain size, and your company may fall into one of these categories as well. Your perception of your direct competition may actually be a company that could become an excellent collaborative partner rather than a competitor.

Indirect competitors may be niche industries or offshore suppliers. Companies that decide to develop business in competition with these indirect competitors may end up sending messages to the marketplace (current and prospective customers) that can be interpreted positively (your company is expanding) or negatively (your company has never before bid on these types of projects so you must be desperate for business). Carefully scrutinize all factors before making a decision to compete in these types of markets.

Competitive Advantage

The Competitive Advantage section discusses the reasons you are successful in the market space and industry segments in which you realistically compete. Your Competitive Advantage section needs to be a stated qualification of the value you provide to your existing marketplace and client base. Compare these value criteria with the Voice of the Customer requirements you cited in your Customer Analysis and the value propositions you developed for yourself. If the value you provide matches your current and future customer requirements, your company is poised for expansion and growth.

If you are planning to expand your market penetration into a niche requiring high quality with tight tolerances, yet yours is a company known for being a rapid-turnaround generalist, you may be misfocused based on your current core business. Perhaps your competitive advantage is your ability to offer great turnkey response in a niche market in which larger original equipment manufacturers (OEMs) or service companies cannot operate profitably. Your competitive advantage may be a matter of your location, logistics, or your company's attention to employee training and certification. Be realistic about your competitive advantage, and avoid boilerplate language like "price-value-quality." If you use these words, use a value proposition to quantify the specifics of how you deliver on each one. Don't count on the reviewer to fill in the blanks and come to the same conclusions you have about your company. Words are powerful. Be specific.

Section 7: Marketing and Sales Plan

"What we really need is a mindset shift that will make us relevant to today's consumers, a mindset shift from 'telling and selling' to building relationships."
- Jim Stengel, Former Global Marketing Officer, Procter & Gamble

There are three or more elements to your Marketing and Sales Plan section. They should address (a) how you plan to attract new customers and maintain current customers, (b) how your sales strategy is implemented, and (c) how you structure your company's pricing, turnaround, and delivery of products, services, and other deliverables. Many B2B companies perceive the discipline of marketing as the equivalent of sprinkles on a cupcake. I have news for you; your marketing plan is the front end of the cash-flow cycle. Marketing *is* part of the total cupcake. Forget about the sprinkles. There is nothing superfluous about the marketing and sales functions in a well-integrated, team-based company infrastructure.

Many technically oriented businesses operate under the misconception that everyone knows who they are (word-of-mouth syndrome); yet they don't have all the business there is to win in spite of everyone apparently knowing who they are. While everyone may know about your company, unless you regularly monitor your online brand and image and engage in Voice of the Customer polling of your customers, you may not realize that you are not well-regarded in the marketplace. So while everyone may know who you are, that may not mean good news.

A well-structured business plan is implemented via a well-structured Marketing and Sales Plan. Most companies, particularly small to mid-sized ones in the industrial and technical service communities, regard tactical marketing (marketing communications) as the equivalent of throwing cooked spaghetti against the wall from time to time and hoping it sticks. These companies are in and out of the eye of their constituents. They will market for a few years, have a booth at a trade show, or engage in online marketing programs, only to cease these efforts the next year because of their perception that these marketing efforts were ineffective. Consistency in your efforts is the key to marketing success.

Marketing Communications Plan

Marketing communications reach out to your target audiences to provide messages about *what* they are looking for, *where* they are looking, and *when* they are looking via strong and consistent marketing statements. Marketing communications is the tactical implementation of strategic marketing initiatives, and it is used to attract new markets and customers. Marketing communications is a commitment, not a hobby, and your Marketing Communications Plan shows your understanding of this tenet.

Based on your annual Industry Analysis and your understanding of your customer base and your Competitive Analysis, you should have a finger constantly on the pulse of what makes your business tick. Determine the best marketing initiatives for supporting the efforts of your sales and business development professionals. This insight becomes critical in keeping your company top of mind – in front of not only current customers, but also new customers and emerging markets.

The best salesperson and marketing spokesperson you have for your business is your website. It is your unpaid, 24/7/365-day salesperson operating across all time zones simultaneously. Not having a professionally designed website makes your company look out of touch. If visitors to your website have to struggle to understand how to navigate it, you might as well prevent them from accessing your website in the first place. Keep in mind that English is not everyone's first language, and navigation should be intuitive for anyone visiting your site. Build your website for the visitor, not for yourself. Not everyone visiting your website is in buying mode, so being able to track traffic and pursue contacts, even if they are looking but not specifying, is an important business development strategy.

Other forms of marketing communications include, but are not limited to, targeted advertising on relevant, industry-focused websites; in online trade magazines; and in hard copy journals. You may elect to purchase an Internet marketing program or work with an advertising agency. Doing nothing is not an option, and it certainly isn't effective!

You have to work to earn your business – there is no easy way around it. Learn how to translate your great idea into the language that best expresses the needs of your target markets and your target customers. Then make that message easy and accessible for them. Speak in common denominators, just like you learned to do in meetings with your colleagues.

Sales Plan and Revenue Generation

What does the sales initiative look like for your company? Some companies have their own dedicated sales force, while others elect to hire manufacturers' reps to serve as their hired sales guns. Inside sales is becoming an important area for business development since your very busy customers and prospects can now find out what they need to know about your company and your competitors online. Companies may also have a separate customer service function that supports their sales force and, like inside sales, this function is growing in importance relative to business development and customer retention. Companies can divide up their sales force regionally; others segment by vertical target market. Each of these areas requires consideration and discussion in the Sales Plan and Revenue Generation section of your business plan so that you are not perceived as thrashing around in the marketplace, selling anything you can to anyone you can.

In very small companies, the president may take off her presidential hat, put on a sales hat, and go on the road to sell, return to the company and put on another hat to implement and direct manufacturing on the sold business, and then put on yet another hat to oversee generation of invoicing. Is this what the sales function looks like in your company? Often, in writing this section of the business plan, companies begin to realize that one person (like that president) cannot continue to be all things to all people without negatively impacting business growth. After all, sales cannot come to a grinding halt once you stop selling and start manufacturing. What is the effect on long-term cash flow of continually repeating this process?

Salespeople and business development specialists should have a very clear idea of their target markets and target industries. Your sales organization (even if it is only you!) should have a process for developing relationships, identifying opportunities, and presenting solutions or products. Business development involves researching the systems in which your solution or product is being placed, not just selling the solution as a remedy for a customer "pain point," which may not be accurate. Have a plan and a process, and access many of the excellent books written about sales methods. I provide some resources and references in the back of this book.

Pricing Structure

Your pricing structure for products, services, and other deliverables is included in this section of your business plan. Hopefully you are not making up pricing arbitrarily on a per-customer basis as you go along. In this section, you evaluate whether your products are competitively priced within a commoditized marketplace or whether they represent unique solutions at a premium price. You need to assess your customer base and how your customers affect your pricing structure. If you do business with companies that are constantly forcing you to provide products and services at lower prices regardless of inflation, you need to examine how this situation impacts overall cash flow, revenue generation, and profitability. You may feel good about having a large customer base. But if you have to maintain a large base to compensate for giving products and services away, you may not be moving in the direction in which you want your company to move.

REVIEW OF MAIN POINTS

1. The Industry Analysis section of a business plan provides the rationale for why you want to start or started a business in a particular marketplace, and why you continue to do business there. In this section, you discuss market opportunities and the drivers providing the impetus for your company to develop and retain business.

2. The Market Overview section of the Industry Analysis discusses whether there is a reasonable market for your products, services, and capabilities. How much of the marketplace can you realistically anticipate earning business from?

3. The Customer Analysis section provides an important means of reinforcing why you do business with your current customer base. How did you attract a particular type of customer in the past, and does this type of customer still make sense for your company?

4. The Competitive Analysis section looks at your company's strengths and weaknesses and helps you decide how to capitalize on your strengths within a realistic market space. It evaluates how you present yourself and your capabilities to new and existing clients.

5. A well-structured business plan is implemented via a well-structured marketing and sales plan. Your sales folks should be delivering marketing and sales messages about your technical deliverables that are consistent with your business plan. Your sales efforts should be organized to effectively implement marketing strategies, generate revenue, and consistently keep cash flowing into your company.

6. The Pricing Structure section evaluates whether your products are competitively priced in a commoditized marketplace or whether they represent unique solutions at a premium price. You will need to evaluate your customer base and how your customers affect your pricing structure.

SALES-ENGINEERING INTERFACE™ TOOL # 11: YOUR INDUSTRY, YOUR COMPETITORS, AND YOUR MARKETS

1. Go online and Google *markets*, *competitors*, and *industry analysis*. Is this information news to you? Does your company regularly gather information and send it to all departments? Why or why isn't this information shared? How aware are you of trends in your industry? Subscribe to RSS feeds and blogs related to your professional discipline so you can familiarize yourself with the movers and shakers who impact opinion and technology for your industry.

2. Take a look at your current customer base. Are they a certain personality type or company type? What makes these customers a good fit for your company? What are their needs? Who are your direct and indirect competitors? Why does your company feel it is competitive?

3. What gives your company a competitive advantage in your industry? Do you provide unique products and services or delivery of a technology? Are you a commodity producer? Is your company moving in the direction you wish to move? Why or why not?

4. What type of marketing and sales plan does your company have in place? Which individuals and departments are involved? What are the objectives of your marketing and sales plan? Does this plan remain constant or does it vary annually? What are the reasons for stasis or change?

CHAPTER TWELVE

ACHIEVING YOUR MILESTONES

Section 8: Operations Plan and Key Operational Processes

"Holding back technology to preserve broken business models is like allowing blacksmiths to veto the internal combustion engine in order to protect their horseshoes."
~ Don Tapscott and Anthony D. Williams, *Wikinomics*

The Operations Plan and Key Operational Processes section of your business plan provides a detailed description of how your company produces and delivers output – namely manufactured goods and/or services. You can fine-tune this section to fit your particular business. You will need to address, in detail, the day-to-day operational processes that allow your company to serve its customers. Provide a succinct description of the function of each department and division in your company.

Day-to-Day Operational Plan
This section is a step-by-step listing of the processes and personnel involved in how your company serves its customer base, from order to cash. The reader must be able to understand the steps, departments, and individuals per job function/title involved in obtaining an order, being awarded a contract, and getting that order out the door or project completed, invoiced, and paid for in full. Describe this process, including a brief description of quality processes and certifications and/or registrations you, your company, or your staff hold or plan to obtain. Discuss whether and when additional

contracted staff are needed on a full- or part-time basis in order to implement projects. Components of the Day-to-Day Operational Plan, which focuses on key operational processes, include, but are not limited to:

1. Marketing and sales: new customer acquisition and client retention (a brief summary of the material created for Section 7 of the business plan)

2. Accounts payable and receivable (vendor and client processes)

3. Accounting and payroll (human assets, including permanent and contracted workforces, consultants, benefits, and tax liabilities involved with human assets)

4. Raw materials and supplies purchasing

5. Customer service (and how it complements customer acquisition and retention, and/or whether this function is involved in accounts payable and receivable)

6. Quality control processes, reporting, and documentation

7. Design and custom capabilities

8. Manufacturing, assembly, and production

9. Shipping and receiving

10. Supplier acquisition and qualification (sub-contractor identification and vetting processes)

11. Employee training, including quality aspects

12. Documentation and contracts processes, including format and storage

Section 9: Business Model

The Business Model section of your business plan is where you provide the reader with a snapshot of the structure of your company and the people you have in place – or need to put in place – in order to move forward. The discussion should include, and certainly isn't limited to:

1. The business model under which your company is currently organized, including the major operating divisions and their interactions and interdependencies with each other. Consider whether your business model addresses how your company has grown or will grow – if you are an entrepreneur or a spin-off division of a mature company.

2. A determination about whether your current business model permits you to allocate resources appropriately to target new customers or markets. Will you be hindering the viability of one area of your business model by moving resources (human, financial, and other) to develop new products or create a new division?

3. Tactical human assets responsible for supporting day-to-day operations and how these individuals and functions allow your company to fulfill the milestones and long-term strategies of your business plan. How much overlap is there between departmental or divisional structure and the personnel manning the positions in these departments? Do you have personnel wearing multiple hats and serving multiple functions on an as-needed basis? How does this type of personnel structure within your business model impact the ability to move your company to the next level?

4. Key human asset gaps in your management team (C- ,VP-, and GM-level) that prevent your company from moving forward and achieving its short-term objectives and long-term goals.

5. A list of your board members, if applicable to your business model, and their titles and functions.

Does Your Current Business Model Reflect the Status Quo?

Some companies end up renovating their business model during the business planning process so that it accurately reflects their current business focus and they are positioned for flexibility and growth.

Is your company organized horizontally or vertically? Are there divisional fiefdoms in place (siloed infrastructure), or is there an option for reconfiguration and cross-functionality? Some small to mid-sized companies and sole/small proprietorships function by everyone wearing multiple hats and rolling up their sleeves to do what needs to be done to make the product, sell it, and get it out the door. While this dynamic may be exciting, you should determine whether your current business model is costing your business money and profitability.

In describing their current processes (Section 8), many small to mid-sized companies often realize that day-to-day operations resemble organized chaos. With chaos as the status quo or norm, the company often has lost sight of the value of their leadership in leading rather than rolling up their sleeves and joining in the fray. Many mature businesses, regardless of their size, view their longevity as a testimonial to a great business model and good leadership. However, when re-assessing their business model, they often find that they have lost sight of business goals and opportunities: their entire business is now modeled in support of day-to-day workarounds and dysfunction.

Key Human Assets

Human assets – the key management team members – are an important consideration for your business plan. If you are the leader of your company, and your role involves organized, tactical firefighting on a daily basis, it's going to be difficult for you to achieve long-term strategic goals! Identify gaps in the personnel associated with key functions in your organization. If your key management personnel are being pulled in all sorts of directions, putting out fires and constantly managing oversight and crises, your company won't be able to move to the next level and build its client base. If you are a family-run business, will you need to hire additional key strategic personnel from outside the family to fulfill the long-term goals of your business plan? How does your allegiance to the status quo of human assets impact your company's ability to grow or shift position in the marketplace?

The Value of Board Members

Not all organizations have board members and not all organizations need a board of directors. Some organizations have a group of individuals who meet, pro bono, on a quarterly basis, and serve as a think tank for future growth strategies for the organization. Other companies have a board whose members own a share of the organization and are compensated based on profits earned. Entrepreneurships are often funded by venture capital firms that create an oversight board with return-on-investment expectations. Which model fits the long-term needs of your business plan?

Section 10: Milestones

Key business milestones are the critical short- and long-term objectives your company plans to fulfill over the next twelve to thirty-six months. The Milestones section appears at the end of the document rather than the beginning. If it were included at the beginning of a business plan it would look like wishful thinking on your part. By the time one reads the Milestones discussion, the groundwork and infrastructure to support the defined milestones have been laid out throughout the preceding sections of the document.

If you identify and quantify a milestone at this point that you have no means of supporting within the current document, then it has no reason to be included in the business plan. If the milestone is critical to your success but was not defined until this point in the process, then you need to reexamine its impact on the entire business plan and make appropriate changes to the entire document.

Milestones can include, but are not limited to:

1. Relocation, improvement, or expansion of your current premises of operation

2. Employee training and certification programs

3. Company certification goals

4. Expansion of functional areas of the business model with key management and departmental hires

5. Realignment of functional areas to support the business development strategy

6. Purchase and implementation of equipment and software programs (ERP, warehousing, inventory, etc.) to support the business plan

7. Increases in sales and profitability in some or all divisions and their anticipated impact on overall profitability based on prior performance data

Section 11: Revenue Model

The Revenue Model section of your business plan consists of a high-level discussion of the structure, nature, and deliverables of your revenue model and the financial highlights or accounting data from the past five years, if available. (This may not be the case for entrepreneurships.)

As discussed in Chapter 10, there are two sections that most reviewers read immediately upon receiving your business plan: the Executive Summary and the Revenue Model. You write your Executive Summary last, after you have completed your entire business plan, including the Revenue

Model. You start to gather the data for the revenue model as you begin writing the main body of the business plan. It all starts to come together in this section.

The Revenue Model section discusses how the divisions of your company contribute to your revenue stream. If you procure and distribute standard components, discuss the contribution (gross income and profitability) of this revenue to the annual volume of sales for this division of your company. Does it make sense to develop a division of your company by out-sourcing your engineering services to offset cyclical shortfalls in other divisions? If you specialize in custom projects that involve long-term timelines to completion and larger-scale deliverables, and ultimately mean higher profit margins for your company, consider how they impact your monthly cash flow. Perhaps you can develop expertise in turnkey projects to create short-term revenue while you wait for your customers to pay you on the bigger projects.

If your company lets business "happen" instead of having a clear-cut strategy for developing business, this section will be instrumental in organizing these efforts more productively and profitably. Companies with under $1 million USD in annual gross revenue (often start-ups, sole proprietorships or limited partnerships), and companies operating in the $1-2.5 million USD category, are so caught up in generating their own paychecks and covering overhead that they are indiscriminant when it comes to business development. They sell a project, take off their sales hats, put on their technical hats to complete and bill the project, and then put their sales hats right back on to try to generate more revenue. The diagnostic aspects of describing the reality of their business development "process" – and the impediment of their current actions on their overall organization and cash flow – are quite a reality check. Often these companies realign, reorganize, or seek strategic partnership with a similarly sized company with complementary core capabilities.

Financial Highlights

The Financial Highlights section can be very enlightening. While the financial discussion and documentation can become very detailed and complex, it doesn't have to be. And please make these data easy to read and understand! This section should include comparative income data from the past five years (or as many years for which you have data) that focus on:

1. Gross revenue (which can include income from grants, awards, or personal finances in the case of entrepreneurs)

2. Gross profit from revenue

3. Net income

4. Percent net profit

Financial highlights also include comparative expenditure/expense data from the past five years, if available, which focus on, at the very least:

1. Total gross revenue (income) from all divisions, broken out per division

2. Net profit, correlated with divisional data

3. Total customers generating gross and net revenue and profit

4. Total customers per division, including divisional overlap

5. Total profitability

When times are good, one tends to ignore adhering to a structured business plan. Your Financial Highlights may show you how some divisions carry the load – and the company – due to high profit margins and low manpower and raw material needs. Some companies are too focused on one division and ignore developing business in other, more potentially profitable divisions. During the last economic meltdown, lack of diversification in how companies were organized and how divisions were utilized had a negative impact on bottom lines.

Your financial highlights will help you understand, and help your company determine, what divisions or departments are the most expensive to run in the short versus long term, and what the profitability is of maintaining these divisions. Your financial highlights will tell you how your company manages to stay in business, whether it is stuck in a static mode, or whether it is nimble, diverse, and ready for growth.

REVIEW OF MAIN POINTS

1. The Operation Plan and Key Operational Processes section of a business plan provides a detailed description of how your company produces and delivers output – manufactured goods and/or services – on a day-to-day basis.

2. Your Business Model assesses how your business is structured and whether it needs to be reorganized and aligned for growth.

3. Define gaps in your management team by looking at the type of personnel currently employed by your company and the functionality needed to move your company to the next level, if that is your goal.

4. The Financial Highlights section helps you and your company determine which divisions are the most expensive and/or the most profitable to run. Your financial highlights tell you how your company manages to stay in business, whether it is stuck in a static mode, or whether it is nimble, diverse, and ready for growth.

SALES-ENGINEERING INTERFACE™ TOOL #12: COMPARING FANTASY WITH REALITY

1. Take a plant tour or an office tour, depending on whether or not you are involved in manufacturing. This is when you connect the dots between all the departments. Who is involved in all the processes in your company from procuring an order to invoicing? Could you create a process map of this throughput?

2. Which operational processes are key to your company? If you don't know, make an appointment to speak with the general manager and individuals responsible for process control. Compare and contrast how they perceive this question. Depending on where they sit around the table, they may see the same things differently!

3. What does your business model look like? Does your company operate according to its original business model or has it changed through the years? What *should* your business model look like? Why?

4. Where and how does your company generate revenue? Do you have a homogeneous or diverse customer base? Who are your biggest customers? What has your corporate profitability looked like in the past five years?

Part Four

BECOMING THE GO-TO PERSON

CHAPTER THIRTEEN

DO YOU SIMULTANEOUSLY SIT ON BOTH SIDES OF THE TABLE?

Put Yourself in Your Customer's Shoes before You Try To Sell Them Anything

Business development is about understanding your current and potential markets and the customers in these markets, and developing products and services to meet their needs. Business development is about how well you know your customers, their mindsets, and the context in which they make decisions. The professionals and companies having the greatest appreciation for business development provide the best value to their companies, their customers, and themselves. They are the least frustrated with their careers and business choices and are better able to leverage these with other companies and new positions. They understand what it feels like to wear their customers' shoes. And they are able to sit, simultaneously, on both sides of the table, understanding the technical and non-technical dialogue and the decisions that need to be made.

Your customers have many things on their minds: poor performance by their company in the last fiscal quarter, a strike in another country impacting the arrival of vital raw materials, the price of fuel, cost increases from third parties, not to mention personal issues like health and family that complicate work activities. Your ability to put yourself in each of your customers' shoes is critical to creating robust, collaborative solutions. It is never all about you and your company's solution. It's always about understanding the context in which your customer is making a decision. No matter how many times your prospect or customer acknowledges their pain and that your solution fits their needs, unless you understand the specific dynamic environment into which your solution must be assimilated, you may never win their business.

Although the sales process is equated with business development, it actually occurs in the middle or towards the end of the business development cycle. The sales process focuses on directing the customer/prospect towards placement of your company's product or solution. Starting the sales process prematurely in the business development cycle assumes that your company provides the only viable solution for your customer or prospect. And while your company may have convinced *you* about how great it is, there is never only one choice or one solution for a customer or prospect. Before you sell your customer anything, make sure you are wearing their shoes.

The business development and sales processes are frustrating! You need to be patient and not give up too soon. The client acquisition cycle can be lengthy. And it's not that easy to replace lost customers. Your customers are not disposable. Avoid becoming complacent based on your accomplishments; customer loyalty is never guaranteed. Vendors, on the other hand, can always be sacrificed by their clients, no matter how much value the vendor feels they provide for their clients. The decision-making environment can change at any time. We can be in one day and out the next. Just read the business sections of the newspapers. When it comes right down to it, even you are disposable unless you are extremely valuable to your company. Work towards earning your internal and external customers' business, daily, by constantly wearing their collective shoes.

Compare the Status Quo and Their Perception of Your Value

Putting yourself in your customer's shoes involves understanding their status quo as well as yours and your company's. We have discussed your own status-quo mindset and corporate culture, and how it biases and impacts the collaborative potential of internal teams. Your customers are dealing with the same types of environments – some better, some worse. Your status quo either conflicts with or complements that of your customers and prospects.

Unfortunately, most of us assume other people make decisions based on how we, ourselves, make the same decisions. We tend to create strategies and technical solutions based on our own personal biases and status quo rather than as a result of developing a thorough understanding of what is going on within our own company and the client's company. Being aware of how we inject our personal perspectives and biases into this dynamic equation is the first step in developing a successful attitude and approach towards business development.

Your customers are constantly assessing the value you provide to their organizations. If you are perceived as merely an order-taker or implementer, then your customers and prospects will draw a mental line in terms of how much information they choose to share with you. Order-takers and implementers ("do-ers") are perceived as being shortsighted and tactically oriented. If you are an implementer-salesperson, your customer or prospect often assumes you are only after closing the sale and obtaining their signature on a contract. They may limit you to small sales rather than large projects with more responsibility, more obligation, and a larger commission. If you are a technical professional, clients may perceive you as a cubicle-sitter who, like that waiter at the short-order restaurant, is only interested in rushing off to the kitchen and serving the solution as ordered. And a shortsighted focus, whether technical or not, may affect your company's ability to win long-term, complex projects.

Customers do not equate innovation and creativity with order-taking and implementation. And developing a hybridized technical/non-technical business development perspective is an innovative, professional approach. Your growing core capabilities within the technical/non-technical continuum will allow you to provide greater value for clients and prospects. In this hybridized role, you offer the potential for becoming that partner on your customer's decision-making team.

Clients Avoid Making Decisions

Start applying your cross-functional, technical/non-technical business development perspective to your clients' decision-making processes. Think about your own household and how long you wait until you replace appliances or contract for a plumber's services. If you are technically inclined, you are confident you can fix things yourself, whether your self-assessment is realistic or not. If you are non-technical, you may want to rush off to call a repair person, but budgetary considerations force you to at least try to fix things yourself. Your customers' "houses" have the same mindset: first and foremost, they will try to repair faulty processes themselves. Do you know how embarrassing it is for your clients to admit that there is a problem they can't fix using their internal capabilities and highly paid technical personnel?

The business development process starts with your upfront homework of determining what your customers' and prospects' houses look like before you make an unscheduled, or even a scheduled fix-it call, because that's how your customers regard each interaction they have with both technical and non-technical vendors who make appointments or schedule conference calls. Your customers anticipate that you are going to charge into their house and attempt to fix them. That's threatening for your clients. No matter how badly their business pipes are leaking, they still feel that they can apply one more layer of duct tape to keep their business afloat. They will avoid going outside for solutions.

Your role is to build your value and trust by creating collaborative dialogue with clients. Your new, hybridized, technical/non-technical business development approach will give you the patience to continue this dialogue. Refrain from switching gears into engineering or sales mode once your client starts unveiling their internal dysfunction or technical problems. Just listen and resist the urge to fix. Keep in mind that your clients and prospects will fight the inevitable – their need to seek outside solutions. Understand that your taking a strategic perspective may be far more rewarding in the long run than forcing a tactical duct-tape solution onto your customer. Root causes can have really big contexts.

The Sales Process Is Broken

It's difficult to sit back, relax, and have those meaningful discussions with your clients and prospects when you constantly feel pressured by your company to perform. While you are taking the time to develop a hybridized technical/non-technical business development approach, the majority of companies, including yours, may have an outdated perspective of the sales process. If you are in sales you understand this; the slate is always wiped clean at the end of each quarterly sales campaign. This means you are only as good as your last sale, and the company always seems to be asking you, "What have you done for me lately?" If you are a technical professional, you are only as good as the profit margin on your last job and your company's perception of whether you are using up too many non-billable hours to serve your customer base. Companies have short memories.

This sales paradigm is broken. Technical and non-technical professionals end up desperately churning and burning to speak with as many customers and prospects as they can, figuring at least one or two will be "sales ready" or throw them a bone in the form of a request for proposal (RFP). Everyone is so caught up in this mythology that it is difficult to extricate yourself from this mindset. You work with it every day. However, now you are also simultaneously working towards utilizing a cross-functional business development approach. While it may seem for a while like you are trying to rub your stomach and pat your head at the same time, there is a light at the end of this tunnel!

How do you break away from the broken sales models and status-quo habits in your own career? You won't be able to do this all at once, and neither will your customers and prospects. Many of them are in the same boat. Start to chip away at the situation. Identify the prospects and customers with whom you have the best rapport and those who have the greatest appreciation of your personal, hybridized, cross-functional currency. Working with these clients can provide you breathing space and the opportunity to inject the skill set, tools, and perspectives you are learning in this book, one project at a time. Determine which companies and projects might best lend themselves to your moving away from the old status-quo sales-and-engineering model towards a new, hybridized, technical/non-technical direction in which you have greater control and understanding of the business development process.

Develop A-List Customers by Being on Their Go-To List

In sports, some players seem to be thinking three plays ahead of what's currently happening on the field. They anticipate not only the future moves of their teammates, but the actions of members of the opposing team as well. Their foresight allows them to be in the right place at the right time to create opportunities for their team. They take a hybridized approach to the game; they simultaneously think on both sides of the field.

Become that individual for your company and your customers. Develop a 360-degree perspective so you can be proactive and anticipatory, regardless of the siloed mindset that may be swirling around you. Allow yourself to sit, simultaneously, on both sides of the table. The dynamics of taking this type of approach are palpable. It's hard to hunker down into a status-quo, siloed mindset when you are asking the types of questions that make everyone stretch their capabilities. Just as that star player makes everyone on their team a better player, you can catalyze your colleagues to up their own game.

The same can be said for companies seeking to break out of their own status quo and develop new markets and new customers by utilizing a cross-functional, technical/non-technical business development approach. Each chapter of this book, although directed to the individual, also speaks to your company and C-level management as a whole. You can't move forward until you understand what is holding you back. Often an outdated business model and a status-quo customer base create a lot of elephants impeding forward progress. Take the time, as a professional and as a company, to assess who your best customer is and determine what makes them your best customer. You may end up pointing yourself in a new and robust direction.

Client mix, or what your customer base looks like, is important. Some companies feel a good customer is one who provides them with the highest volume of repeat business during the year. Others feel a good customer is the one who brings them high-level engineering, custom design, and fabrication work. Most companies have built their business base on a mixed bag of customers. When they started out, they were hungry for business. They did business with any customer who would give them work! Now they don't have the heart to make that important shift to focusing on the specific types of clients who will take their businesses to the next level. These companies can't seem to cut loose those customers who were there from the start. They need to break away from status-quo clients, just as you are gradually breaking away from that status-quo mindset.

Most companies do not take the time to focus on how collaboration with a customer can make that customer their best customer. Collaborative, cross-functional relationships streamline decision-making and facilitate communication. Collaboration shortens the business development cycle because everyone inherently tends to be on the same page. Differences are hashed out expediently. And the order-to-cash cycle is shortened, enhancing cash flow, revenue generation, and profitability.

There's a tremendous upside to having a hybridized, cross-functional, technical/non-technical perspective for business development. You may have a mixed bag of customers for your client base, but if all of them thrive on collaborative relationships, think about how much more streamlined, efficient, and enjoyable those business relationships might be. You would be doing business with your best customer the majority of the time. And you could choose which companies to do business with based on your specifications for a best customer.

Consider what would happen to your business, your professional development, and your bottom line if all of your customers were like your best customer and you were their go-to trusted supplier and advisor. Those are the common denominators you should seek across your client mix. You don't need more customers; you need more customers like the ones you do your best work for.

REVIEW OF MAIN POINTS

1. Business development is all about how well you know your customers, their mindsets, and the context of their decisions. Professionals having the greatest appreciation for this aspect of business development provide the greatest value to their companies and to themselves.

2. There is never one root cause or reproducible formula for business acquisition. Your customers alone know why – and why not – they make decisions the way they do. And they don't make decisions the same way you do. Take the time to discover the factors that can impact or derail their decisions to do business with you and your company. Put yourself in your customers' shoes.

3. The sales process is broken. It assumes that if you can identify a need for your product or service in a potential client, they will decide to place your solution. Your customers don't want you to lead them down this path. They don't need to be fixed. They've got bigger factors to deal with. Your solution may not be at the top of their to-do lists.

4. Companies should consider working primarily with customers who fit the profile of the best type of company for their core capabilities and output. Instead of trying to be all things to all people, consider what it would take to make your company the best in class for specific types of products and solutions. What would the economic impact on your company be if the majority of your customers fell into the best customer category?

SALES-ENGINEERING INTERFACE™ TOOL #13: BE THE BEST AND WORK WITH THE BEST

1. Does business development for your company look like endless responses to requests-for-quotes and endless proposal generation, or is it a result of relationship-building, cross-functional knowledge management, and the resulting long-term, multiple-phase projects? Give examples.

2. What skill set is required for you to target best customers? How many individuals in your organization currently have this cross-functional, hybridized skill set? Do you?

3. Who are your best customers? What makes them the "best" in terms of quality, quantity, profitability, or cross-functional collaboration?

4. How could you shift the greatest percentage of your customer base to best customer status within the next six to twelve months? What would the impact be on personnel, equipment, and, once again, cash flow?

CHAPTER FOURTEEN

THE ART OF BUSINESS DEVELOPMENT

Did *You* Ever Learn How To Develop Business?

"Traditional methods of sales prospecting are grossly inefficient."
~ Jill Konrath, *SNAP Selling: Speed Up Sales and Win More Business with Today's Frazzled Customers*

Pry your technical fingers off the ledge your entrenched habits of expertise are clinging to. Technical professionals need to transition into taking a more active role in the business development process. Sales and marketing professionals should give themselves permission to turn on the other half of their brains so they can productively participate in the technical aspects of the business development process. Come on. You both can do it.

Regardless of whether you are technically or non-technically inclined, you have been developing and selling yourself for years. How else did you gain acceptance into a university or technical school? That letter of acceptance didn't just arrive in the mail on its own. You heard (aka prospected) about an institution or training facility, researched that facility (performed business intelligence research), sat for college entrance exams (certified yourself), applied for admission – including writing the admissions essay (marketed yourself), and were accepted for admission (closed the deal). You interested an institution in investing their time, money, and materials in your outcome. That sounds like the business development cycle to me.

Business development incorporates many of these elements. Developing the technical expertise and customer relationships necessary for customer retention is an extension of who you are. You have a compelling story to tell, and who wouldn't want to listen? For that matter, your potential or current customer also has a compelling story. It's your ability to listen, and keep listening, instead of rushing off to the drawing board or closing a small sale prematurely, that impacts your success at developing business.

Business development is not achieved in a vacuum. Taking the time to develop the business case for your customer – a 360-degree perspective of the factors involved in their decision to specify your company – allows you to have your eyes and ears open for factors that impact project outcome and scope. And that 360-degree perspective hinges on your having a technical/non-technical business development perspective. A business case involves more than determining the customer's pain points and needs and then fitting their square pegs into the round holes of your company's solutions. The nature of business development involves patient listening, which can be daunting to those of you who are anxious to demonstrate your acumen by immediately rushing off to solve what you heard was the problem. Root causes can have very large contexts. You must identify as large a context as possible for the problem at hand. Your foresight will result in robust solutions and fewer issues cropping up later to derail the project.

Are Sales Funnels Another Version of Turning Your Company into an RFP Mill?

With today's economic and competitive challenges, all members of a company will be called upon to participate in the business development process for their organization. A bad experience with a customer service representative can cause a client to go elsewhere. An administrator who does not know to whom to direct inquiries creates a poor company image. An invoicing issue that remains unresolved may compound a bad feeling your customer's accounting department picked up about your company from an initial honest mistake. Everyone counts on your team and everyone is part of your team.

Today's technical staff needs to be aware of how critical their function is to the business development process. Your technical manager (or you) is likely responsible for cost containment measures impacting your company's bottom line. Although some companies engage the services of sales engineers or have manufacturer's reps handling the sales function, what will you do, as an engineer, if a customer calls you directly and identifies a new project? If you hand their phone call or email off to the sales or marketing people (if you even have any), how certain can you be that these individuals will respond to the customer query in a timely or appropriate manner? If the customer emailed or phoned you directly, there may be an urgent technical need. Their momentum and interest may be diminished if you pass them off to someone who may not be available or knowledgeable, which can result in a lost sale.

That being said, not all current or potential customers who call you have an urgent need. In fact, they may just have too much time on their hands and want to kick around a concept on your non-billable time. You may be giving away information, thinking you are selling, when you are actually leaking expertise for free to an individual with no interest in doing business with you. How can you tell the difference?

Salespeople (and engineers who also sell) often fall into the trap of turning their companies into RFP (request for proposal) mills simply because they engage in conversations with anyone who will give them an appointment, even non-decision-makers. Salespeople have quotas to fill and often are encouraged to follow a formula that looks like: Make W prospecting calls per week resulting in X appointments, so you can document this information to your boss, issue Y number of proposals to meet that quota, and estimate securing Z number of contracts resulting from these efforts. It's all a numbers game in a broken sales paradigm, right? They equate proposal writing with having a successful selling process. Yet they aren't closing many sales using this method. It's just a lot of busy work.

Engineers in small companies are often asked to churn and burn in the same manner. Their approach involves contacting people they know at other companies and asking if there are any upcoming RFQs (requests for quote) they can bid on. These technical professionals usually have very little sales training in preparation for taking a sales role. Like those sales professionals who are gung-ho to prepare proposals, the engineer may perceive that being asked to create a quote is the objective of the sales process, while most of the time their company is simply being used as the source of the third bid on a project. Somewhere in this "selling" activity is a sales engineer who is brought into the equation too late and spends a lot of time preparing a proposal for a prospect who has no intention of doing business with their company.

> *Have you ever considered how easy it is to get a salesperson out of your office or an engineer off the phone if you tell them to send you a proposal? Isn't your professional time worth more than participating in dead-end busy work?*

Have C-level Discussions with Your Customers

In order to develop business, you have to establish a dialogue with your customers. You have to start somewhere, so make it easy on yourself.

Provide Relevant Information

Some of the most comfortable conversations you can have are with your current customers. Often you have access to the CEO or a key decision-maker within the organization. Start practicing your growing skill set by talking business with these valuable resources and, in turn, become a resource for them. Not every conversation you have with current customers needs to involve selling something. However, these decision-makers do appreciate your keeping an eye out for relevant information pertaining to their industry or market segment.

In Chapter 2 we discussed how it is easier for someone to say no when professionals are not providing them with the information they need to say yes. Become an information resource for these very busy decision-makers.

Instead of being perceived only as a solutions provider looking for a sale and a signed contract, you will become an individual who provides go-to value.

Call People You Know or Referrals

Today's very busy decision-makers respond more favorably to engagement with referred vendors than they do to someone who obviously is churning and burning their way through a cold-calling leads list. Respect this trend. And while you may want to practice your growing business development skill set on people you already know – your current customers – be respectful of their time constraints. Key decision-makers, in particular, do not want their time wasted. Most of them don't answer their phones directly, preferring to have their administrator screen for what is of greatest value or importance to their time. Most of the time your phone calls will not get returned unless that individual knows who you are or has heard about you or your company.

If possible, only call on companies when you have a referral. If you have an internal contact, perhaps a peer engineer or marketing colleague with whom you feel comfortable, ask them who the key decision-maker is for their organization and whether you can use your contact's name as the source of referral when calling this decision-maker. Your LinkedIn contacts are a good source of internal referrals, providing you are not an indiscriminant social networker and actually have an internal resource with whom you've had some sort of relational contact. Referrals are valuable when it comes to business development!

Leave Provocative Voicemails

When you make your initial phone call to that decision-maker, you may be sent to voicemail. Don't take this as a sign of disinterest. These people have perhaps one hour per workweek to make personal contact with all the people who want to speak with them. Reinforce your referral status in your voicemail message, and let them hear what your voice sounds like. Leave them that relevant message related to their business – not to your company

– that will make them think about the ideas you are presenting. Pique their interest. Jill Konrath's books, *Selling to BIG Companies* and *SNAP Selling: Speed Up Sales and Win More Business with Today's Frazzled Customers*, are excellent resources for how to gain access to busy decision-makers.

Follow Up

Then follow up. Make sure you do what you say you are going to do in your voicemail messages and emails. Your voicemail message differentiates you from individuals who are simply asking for an appointment or the opportunity to respond to an RFP. Being proposal-hungry demeans your expertise and makes you look like you are desperate for business.

When you follow up with the decision-maker, continue the discussion you started in your voicemail message. Ask some initial questions to show your interest in them, their company, or a recent article you read about them. Asking to close a sale before you have engaged in the business development process makes you look amateurish. Keep in mind that you only have one opportunity to make a great first impression.

Ask Good Questions, Then Listen

Be prepared to walk your talk by *listening*. It's very productive to ask good questions and let the decision-maker do the talking. And while you may be more comfortable speaking to an internal contact somewhere in that organization's food chain, you can discover far more information by conversing with the key decision-maker. However, your discussion must be appropriate and relevant.

Good questions start by putting yourself in the CEO's shoes and asking open-ended questions – those that do not result in yes, no, or maybe answers. Keep in mind that you should have a pretty good idea of the type of conversation you want to initiate as a result of doing background research on the company, its profile, and its business track record.

Asking good questions means varying your method of asking them. Far too many of us start our questions with the word *why*. It's a bad habit. Eliminate this word from your pattern of questioning and you will get richer answers. Constantly asking *why* questions may put the individual on the defensive, as though you are questioning their judgment before they even know you. Using *what*, *where*, *when*, and *how* to frame your questions can actually make individuals think about their responses differently. If you don't want to get status-quo answers, stop asking status-quo questions.

Avoid the Urge to Solve and Sell
Keep in mind that not every discussion will result in a project to bid on or a proposal to prepare. In fact, that should not be your main objective. How many sales professionals have earned the confidence of the key decision-maker by having a high-level discussion, and then blown it by switching into sales-spiel mode or asking for an RFP? Interest in dialogue does not mean interest in hiring you or purchasing your solution. Do not rush the process and derail all of your great client engagement work. The minute the conversation starts to sound self-serving, you've lost the opportunity to be invited to make the sale.

Wear C-Level Shoes
If you are going to engage at the C-level, learn how to have a C-level discussion. In Chapter 9 we discussed the traits of Parinello's VITO™ (that Very Important Top Officer) and how you are the CEO of you and your career. This discussion eventually becomes a peer-level one.

You have a certain amount of expertise to offer these individuals, and they have a certain amount of strategic and contextual insight to teach you. Earn their respect for your approach to business development. Allow yourself be invited to move into the sales process. Often the decision-maker will choose to switch the conversation to the sale without your having to worry about when it is appropriate to do so. They will start asking you about your solution when they feel it is time. Have confidence in your ability to earn their trust. You will find that you gradually become more comfortable with the pace of the business development process.

Okay, So I Have To Develop Business. Now What?

Increasing the scope of work acquired from your existing customer base is an excellent means of business development. Conversing with your existing customers is far more comfortable than calling on companies and individuals you don't already know! Often your existing customers have pigeonholed your company as a provider of specific and limited solutions. They may be unaware of the full scope of services your company can provide. Refrain from calling them up and saying, "Hey, did you know our company can also do this for you?" and letting your customer try to figure out how they can fit your capability into their needs. They won't take the time to do your thinking for you. Instead, provide the context for their decision by starting the conversation with: "I've been researching your company's need for X technology, which may fit in well with our Y capability, resulting in Z increase in productivity for you. Would your team like to develop this discussion?" And let them respond: "Gee, I didn't realize your company could do that." One conversation is a deal-killer, the other one a relationship-builder.

On the other hand, when initiating contact with prospective customers, you're venturing into uncharted waters. You may not be as confident as you are with your current customers. Everyone, salespeople included, tends to feel more vulnerable with prospects. So there is a greater tendency when prospecting, even if referred, to sound like that babbling talking head from Chapters 4, 5, and 6 who sounds like a walking advertisement and is only interested in cold-call prospecting and making an appointment. Anticipate this situation; have information to share even if you feel you are losing your high-level discussion focus. Remember, you are starting a dialogue, not delivering a speech.

Use the resources from Sam Richter's book *Take the Cold Out of Cold Calling* to assemble an initial profile for a potential client and project. As you learned in Chapter 5, sometimes an individual's title may mislead you into assuming they are unimportant, when in fact they have bottom-line responsibility. Some additional areas to focus on when structuring your conversation include:

Use Trigger Events as Talking Points

Look for "trigger events" in industry journals and newspapers. Did one company acquire another? Did someone get a big promotion? Did a company diversify into a new market? Was a capital purchase for equipment made that now qualifies your company to provide services for your prospective customer?

Discover the Business Case

During your conversation, establish the business case for why you contacted them. Congratulate them on their promotion or whatever trigger event prompted your reaching out to them. Have a relevant piece of information to share with them that does not sound like sales spiel or self-promotion. ("I read this article in XYZ Journal and thought it might be relevant for your company.")

Title, Function, and Sense of Urgency

If a prospective customer calls you on the phone and you pick up before you can screen them, qualify the role the individual plays in their organization. Don't be afraid to ask them their title and responsibilities, even if you find out they are the CEO! Are you having a peer discussion with an individual who has time on their hands?

Ask this person about (a) the urgency of the project (keeping in mind that ALL projects seem urgent), (b) the projected timeline of the project (as another means of validating urgency), (c) the scope of the "urgent" project (another way to determine whether the customer has reasonable and feasible objectives), and (d) if the project they are speaking about has been approved and funded! Your time is valuable. Just because someone calls you doesn't mean they are ready to buy or specify.

Document Your Discussion

Take notes so you can discuss your conversation with a potential team member from your company. Your notes should include complete contact information, including title, as well as some additional research on the company and what their track record is all about. Business isn't developed in a vacuum. When I discuss these types of phone calls with my colleagues, I find that each person seated around the table hears the conversation differently. The result is a more complete perspective on the subjects discussed and the opportunity for business development.

If you want to create a meaningful and productive business development pipeline instead of an RFP mill, engage your prospects' top-level decision-makers in your initial conversations. This critical aspect of business development may be outside your current comfort level. There are individuals in your organization who can help you cross-train your perspective and let you sit or listen in on meetings and conversations that demonstrate the dynamics of productive business development. Keep in mind that you already have the knowledge base to engage in a high-level conversation. And you are working hard to develop a technical/non-technical perspective for the business development process. You just need to put what you are learning into practice.

REVIEW OF MAIN POINTS

1. Most technical professionals are very uncomfortable with becoming actively engaged in the business development and sales processes. In fact, whether you are a technical or non-technical professional, you are engaged in the business development and customer retention processes every day.

2. Listen productively to the conversations you have with your peers and decision-makers. The context in which a potential project is defined is often larger than the project itself. There may be opportunity for significant repeat, long-term business. Often technical professionals are in the best position to be involved in these conversations, yet they have not developed the listening skills for recognizing business development opportunities.

3. Resist the temptation to solve problems and sell solutions. Many prospective and current customers may call you hoping you are hungry for business and a potential sale. Don't leak information to a prospect who has no intention of giving you a contract. Take the time to establish a legitimate business case with the key decision-maker.

4. Business development often involves staying in touch with existing customers at the C-level. Provide them with industry-relevant and technologically relevant information; help them grow their businesses. That way they will come to regard you as a go-to resource rather than as an individual who is only interested in closing a sale.

SALES-ENGINEERING INTERFACE™ TOOL #14: GET OUT THERE AND DO IT!

1. Your assignment is to call five of your current customers as a means of keeping your name and your company's name in front of them. This is a customer-retention exercise. Before you contact these individuals (and they can be your peers for comfort level at this point), do your homework. Find out if their company has been in the news, had a new hiring, is involved in a new technology, or has been awarded a new contract. Identify a trigger event that you can use as the reason for your phone call. And this needs to be a *phone call*, not a series of back-and-forth emails.

2. Call your contact at the company. You will already have created a set of ten questions you would like answered. These questions *cannot* begin with the word *why*. Take notes on the answers they provide during your phone call. If your contact asks you whether this is a sales call, be honest and tell them you are gathering and exchanging information so you can stay on top of issues that are relevant to your business relationship and to keeping them competitive. See if they relax once they realize you are not fishing around for an RFP.

3. After your phone call is completed (each phone call should take at least ten minutes and possibly up to thirty minutes if things click), analyze your notes to determine the quality and quantity of information you obtained. Compare whether this information is the same or different from the type of information you have gleaned from prior conversations.

4. If you feel you have uncovered potential opportunities, call in a colleague (perhaps your collaborative partner from the engineering or sales discipline), and discuss your conversation with them. Find out whether they *see* things the same way you *heard* them.

CHAPTER FIFTEEN

YOU ARE THE PHYSICAL EMBODIMENT OF YOUR COMPANY

"Personal branding describes the process by which individuals and entrepreneurs differentiate themselves and stand out from a crowd by identifying and articulating their unique value proposition, whether professional or personal, and then leverage it across platforms with a consistent message or image to achieve a specific goal. In this way, individuals can enhance their recognition as experts in their field, establish reputation and credibility, advance their careers, and build self-confidence."
~ Dan Schawbel, *Me 2.0*

Personal Branding Isn't Just for Them – It's for You, Too

It's time for me once again to pry your fingers away from your biases against using the Internet for anything other than emails. People aren't "buying" your company; they're basing their choice of vendor on your ability to catalyze their decision-making. You are the physical embodiment of their customer experience, whether you are an owner, technical professional, non-technical professional, or the meat counter clerk at the neighborhood grocery store. They are choosing whether or not to do business with YOU.

Business does not happen in a vacuum. And the Internet has created so much transparency that you have the option of managing the information about you and your company or ignoring it. But it still exists, regardless of whether or not you are in denial. Your current customers, potential customers, and competitors use the Internet to find out about you and your company. You can utilize the wealth of information on social media sites to assess industry trends, identify trigger events impacting companies you are targeting for business development, and learn about technological advances that might benefit these companies' business strategies.

Your connections on social media sites and online discussion forums provide a level of primary information and insight that most companies pay marketing research firms large sums to obtain. Peer-reviewed journals are available via online libraries. Everyone benefits in this dynamic forum. If you want to become a thought leader and go-to person, you cannot shy away from this aspect of cross-functional, technical/non-technical business development.

The concept of having a *personal brand* was created by Tom Peters in the 1990s and extended into the digital millennium by Dan Schawbel. While you may feel this is a trendy term limited to Gen-Y/Millenials, guess again. What impression are you communicating about your research approach, the interpretation of your data, and the type of team you have assembled by having a blog, newsletter or website, or publishing research or white papers and making them available online? The impression you create is nothing less than your personal and professional brand – what you bring to your profession based on your values as an individual, a consultant, and a researcher, whether academic or in industry.

While you may feel that personal branding is beneath your intellectual pursuits, your company or institution doesn't. Just ask the marketing, business development, and alumni departments of your academic institution about the importance of personal branding.

If you are interested in attracting graduate students to your institution or employees to your company, guess where they are going to find information? Online. Are you, your company, your institution, or your department missing from their search results? The online perceptions of these potential grad students or employees may very well form the reality of why they apply for acceptance into post-graduate studies or employment (or why they don't!).

Creating a compelling personal and professional brand is not just for job seekers. You can develop the leadership expertise that may be missing from your current career. Use personal branding to gain recognition so you can build towards your next career, attain a stewardship path, or move into retirement mode. Personal branding is a way of differentiating yourself from the crowd. If you are waiting for someone like your boss to notice you and pick you out from everyone else, don't bet on it.

You must market yourself, your skills, and your value in order to be rewarded. Personal branding is not a form of narcissism; it has become a necessity for professional advancement in today's global community.

Personal Branding and Technical Professionals

The concept of personal branding still tends to be perceived as counterintuitive to the industrial sector and the technical mindset. The industrial/manufacturing sector lags behind B2C companies in how they perceive the impact of the Internet in driving opinion, perception, and therefore revenue for their businesses. Professional social networking sites are often viewed only as interactive job boards rather than forums for idea exchange. Technical professionals may be at a loss as to how to keep their names in front of the marketplace while they are currently employed. What happens when you need references for employment or a collaborator for a new technology or business plan? If you wait to develop a personal brand until you need it, you will be too late. A personal branding strategy is proactive, not reactive.

Overcoming the self-articulation hurdle is a tall order – especially for technical professionals – even within the safe confines of the workplace, let alone cyberspace. This hurdle may also limit your ability to be successful in your career. Telling a technical professional that they need to be who they authentically are, especially online, is like speaking Martian. It's time to learn that language. It is not beyond your grasp.

Articulating Your Core Values Online

You've been working on the Sales-Engineering Interface™ exercises throughout this book, and have been building your core value and core capability profile. You will articulate it as part of your resume. That resume includes establishing your profile for your online personal brand. Prospective employers are searching within online communities for articulate individuals with emotional and intellectual intelligence.

Do you communicate with confidence and expertise, and with the ability to lead rather than follow? Do you participate in the business development process? In a recent poll conducted by CareerBuilder, 71 percent of hiring managers (n=2600) said they valued emotional intelligence (EI) above intelligence quotient (IQ). And 75 percent of these managers said they would be more likely to promote an individual with high EI than one with high IQ. (http://www.careerbuilder.com Press Release, August 18, 2011) That is a sobering thought for anyone who feels they are on a higher intellectual plane than the rest of the employees. Are they creating their own me-versus-them silo?

The Four Attributes of Personal Branding

Personal branding allows you to integrate your core capabilities and your core values into your professional profile. Personal branding, according to Dan Schawbel in *Me 2.0*, involves *authenticity, transparency, value,* and *visibility*. I had the opportunity of interviewing Dan Schawbel on August 23, 2011, and created a three-part blog series with audio download regarding our discussion. You can read it at *http://salesaerobicsforengineersblog.com/2011/08/24/personal-branding-and-the-technical-professional-an-interview-with-dan-schawbel—part-1/*. The shortlink is http://bit.ly/n38CsT.

Authenticity is the key to personal branding. It incorporates being the real you consistently across all levels of interaction. It is grounded in your consistent delivery of your core personal and professional values – you aren't switching gears or identities daily or weekly to cater to the opinions of others. Your readers, employers, customers, and students know what they are going to find, whether they are on a social media site or reading a professional journal.

Transparency is about being open and honest. Be who you are, honestly, ethically, and legitimately. Don't copy someone else, plagiarize their work, or attempt to degrade them for your gain. While you feel you may be creating a personal brand for yourself through such efforts, your actions speak louder than your words. Attempting to build yourself up at the expense of others is not perceived positively.

Value involves how you differentiate yourself from the marketplace. All technical professionals are not alike, and neither are all sales and marketing professionals. This entire book is devoted to identifying and articulating those technical/non-technical core values that deliver value to yourself, your organization, and your clients.

Visibility involves how you make yourself, your values, and your credentials known to your audiences. And you do have audiences, whether you admit it or not. If people cannot find information about you online, you are invisible. That is the paradigm of the digital millennium.

You are the physical embodiment of your corporate brand as well – the brand of the company you work for – whether it's your own business or someone else's. With so much business being done virtually, a customer's only point of contact with your corporation may be you. Your corporation's brand becomes synonymous with your service delivery, that customer's experience of working with you, and therefore your personal brand. In this way, personal branding becomes critical to customer retention.

Sometimes an individual's personal brand and way of doing business is stronger than their corporate brand and business plan. Has your company won new business because an engineer from Company A moved to Company B, retaining your services because of the value you brought to them at their former place of employment? Considering the length of the business development cycle, your personal brand may have a lot more to do with your company's revenue stream than you think.

Your Personal Brand Is Your Business Plan

Personal branding involves developing a business plan for *you*. You are the CEO of *you*. You need to develop solid goals for creating your personal brand, whether you want to move jobs, be a thought leader, start a business, or write a book. Just as a business plan provides a consistent set of messages and concepts by which a business is organized, your personal brand should be just as robust. Your efforts need to focus on giving people a compelling reason to do business with you. And in order to do business with you, they have to hire you, don't they?

Your personal brand plan should:

1. reinforce your value to yourself, your customers, and your current and/or future organization

2. create value for career-building and fundraising

3. create an accurate and accountable legacy for yourself and your family (your brand can enhance the marketability of family members, too)

4. enhance the brand of your company or enterprise

5. guide the quality and quantity of information about yourself that appears on the Internet

6. allow you to move in a new direction if you wish, pre or post retirement

You Receive as Much as You Put into Personal Branding

The major platforms for personal branding in the digital millennium are social networking sites and online forums. Personal branding strategies get your brand and message out to the various online communities and user groups sharing or exploring common interests, activities, products, and services. You get as much out of social networking as you are willing to put into it. It is the leading way to land a new job. You are creating demand for your personal brand rather than participating in the flood of resumes sent in response to some online job board, where you are a commodity. And you need to work to maintain your online personal brand. It is not a one-and-done effort, although some social media venues are more low maintenance than others, as I will discuss.

Which social networking venues are best for achieving your goals? You can leap right into social networking venues and thrash away, but keep in mind that the Internet is unforgiving. It's like that tattoo you got during spring break in Panama City while you were in high school. You know, the one you thought your mother would never see? Decide which social networking sites are best for your personal branding goals and always consider what you post online very carefully.

In developing strategies for entering the social networking arena, think about what you want to achieve over the long haul. Revisit Part Three of this book and look at the components of a business plan; these need to be considered when creating a personal brand as well. Here are some suggestions for the questions you should ask yourself before you dive into social networking and personal branding. Decide how they fit into your business plan for your personal brand.

Do I Want To Read but Not Participate?

Do you feel more comfortable using social media sites for reading about topics than for participating in discussions? If so, then www.alltop.com is a blog site that allows you to search by topic for blogs that may be of interest to you. (And if you become more involved, it's a great site for determining which blogs are complementary to your areas of interest.) Read blogs and sign up to receive them via RSS feed. Get a sense of style, who is commenting on what topics, whether the topics are of interest to you, whether you have expertise in the area, and whether you want to remain an active or passive participant. This strategy has low return on investment (ROI), allowing you to dabble with social networking without much associated personal risk.

Do I Want To Locate Groups?
Do you want to seek out specific Internet and non-Internet groups? Again, Alltop is a great place to find out about groups related to blogs. Most of these groups tend to organize themselves on professional networking sites like LinkedIn (www.linkedin.com); and there are tons of others, like Squidoo, Spoke, and Google+ Profiles, to name a few. You can join an online discussion forum and elect to be active or passive in terms of your participation and comfort level. This strategy has moderate to high ROI potential depending on your level of interest and participation. It takes you from a spectator role to that of a participant. And some of these groups may have local chapters where you can meet people in person.

Am I a Blogger?
Do you want to publish a blog to help you establish your personal brand? If so, then you need to make blogging a part of your personal branding business plan and develop your strategy and focus. Once you start blogging, people will start to check you out online. If they Google your name, your profile needs to start coming up on professional sites such as LinkedIn and Google+ or else your blog will not have credibility.

A blog allows you to create a well-managed digital footprint and strong value propositions on social networking sites. Again, if you wish to write a blog, you'd better have a plan that you stick to and a selection of topics that you can speak to. You can go online and download blogging templates to keep you focused and consistent. This is a high-risk strategy. There is a high rate of blog abandonment, and you don't want to drop out just when you are developing a loyal following. Monitor readership via tracking tools like Google Analytics.

Do I Want or Need a Website?

Creating a website to showcase your name, products, and services clearly communicates the value propositions of your personal brand strategy. In our interview, Dan Schawbel recommended that one should own their own name as their domain name. Creating a one-page (at minimum) site for your resume stakes your claim on the Internet and prevents identity theft by companies who buy URLs (and names) and sell them back to people for high prices. What a thought, but true! Other individuals own multiple domain names including their name and key words relevant to their personal brand. At least purchase your own name and populate the site with your resume, which enhances returns of your name in Google searches. Build out your website as you develop your personal brand strategy and professional path. Perhaps you will decide to make a living from marketing and selling the products and services affiliated with your personal brand.

Is There a Book in Me?

Do you want to write a book (hard copy, soft copy, Kindle, or e-reader format) for greater communication and maintenance of your personal brand? Writing a book and having a website and domain name associated with your book is an excellent way to establish yourself as a thought leader and gain credibility with your target markets and peers. It's also time-consuming, and requires some upfront capital. If a book is in your future, where does it occur along your business plan timeline? Books are typically offered in conjunction with webinars as a means of monetizing your personal brand and business. As you can guess, this strategy is high risk, high involvement, and high ROI.

Even if your activities only involve having a strong, updated, current LinkedIn Profile and interacting on the blogs of thought leaders, the larger your Internet/digital footprint, the greater your online brand potential. The greater your online brand potential, the greater the opportunity for dissemination of information and adoption of your ideas and personal brand.

Starting Social Networking

I'm discussing three major social networking sites for professional brand-building in this section. Google+® has just come out at the writing of this book, so future editions will update this discussion as the site is more broadly adopted by users. I don't work for any of these sites, I don't receive compensation from these sites, and the opinions expressed herein are mine based on peer conversations, trend data, and discussions in professional online forums.

For purposes of where to start, I am focusing on the "big three" sites.

LINKEDIN: www.linkedin.com/in/yourname

LinkedIn® is the number one Internet site for professional and personal branding, job-hunting, and business development. LinkedIn has a free basic service that is more than enough for the majority of users. LinkedIn is a solid starting point for any professional personal branding strategy, especially for technical and sales/marketing professionals.

LinkedIn consists of three key elements that should be fully completed to build your personal brand platform.

1. Your *Profile*, which focuses on your personal brand and value propositions. It also serves as a resume.

2. Your *Network*, which allows you to reach out and contact individuals ("Connections") who may be important to your next career.

3. Your *Company Profile*, where your personal brand is linked back to the company for which you work.

Who uses LinkedIn? Your current employer, prospective employers, human resources departments, your current and prospective business partners, your colleagues, and just about everyone else. LinkedIn is where folks go to check you out. And LinkedIn is highly visible on search engines. If there is no information about you on LinkedIn, or only partial or incomplete information, you are missing out on a great opportunity for personal branding.

You don't have to complete your Profile in one sitting, so take some time each day to gradually complete it. But complete it! Include your avatar – your picture, symbol, or logo that you will use across all social networking profiles for consistency and authenticity. The LinkedIn site walks you through each field. There are books written and websites available about how to complete a LinkedIn Profile. Utilize these best practices and tips to complete your Profile. And keep it updated with new accomplishments, references, publications, even your thoughts!

By participating in LinkedIn discussion groups, either as an observer or participant, you can identify individuals and the companies for which they work. You may want to network with these individuals and ask them about their positions and their companies. Perhaps you will decide to work for one of these companies as well!

- **LinkedIn Connections**

Social networking sites work because each profile includes a network of connections – people with whom you either are affiliated or want to be connected to. These connections, in turn, have their connections, and so on. Social networking sites are great for locating individuals with whom you have had a previous business relationship. Perhaps they have moved on and you've lost touch. Perhaps they've married and changed their last name, and you didn't know this information. LinkedIn provides this type of granularity when seeing who is LinkedIn to whom.

There are four basic strategies for creating your LinkedIn Connections list. Yes, even a Connections list needs a plan, too, and your Connections strategy needs to reinforce your personal brand. Where do your personal branding goals fit in?

1. The "I have more Connections than you do" strategy creates a big, bulky Connections list of sometimes unrelated individuals who in turn

have their own big, bulky lists. LinkedIn frowns on this approach, however. The "LION" acronym on LinkedIn denotes a "LinkedIn Open Networker" – an individual who amasses Connections in this manner. Are you merely a collector?

2. The "I have more work-related Connections than you do" strategy results in a big, bulky Connections list composed entirely of people at your place of work and their affiliated networks of LinkedIn Connections. If you feel that having all these folks in your network makes you look important to your company, that is fine. But how does this relate to your personal brand? If your network only encompasses who you work with, and you are still thinking inside that homogeneous box, how can your ideas expand? What if you get fired or decide to look for another job? Then whom do you contact for referrals?

3. The "Personal Brand-Building 101" strategy includes close friends, colleagues, known thought leaders (only invite those with whom you have established a relationship and whose email address you have), and people you want to keep in the loop as you build your personal brand (including selected corporate contacts). For example, I often use my Connections list as my idea sounding board. I post questions to people as a Group of Connections or individually, depending on their areas of expertise. Refrain from adding thought leaders to your Connections list unless you take the time to develop a professional relationship with them. Thought leaders do not appreciate being part of a Connections list amassed by an individual who simply wants them as a status Connection.

4. The "advanced brand building" strategy involves building a network that has multiple tiers of Connections. These tiers involve new Connections with whom you are beginning to develop relationships, high value connections with whom you constantly communicate and collaborate, and thought leaders with whom you've already established

relationships. I involve my high-level and thought-leader Connections for introductions, I comment on their blogs, and I have become a known entity to them as someone who is willing to contribute to their professional success.

- **Joining Groups on LinkedIn**

There are hundreds of thousands of LinkedIn Groups. People searching your Profile can understand more about who you are by the Groups with whom you are affiliated. The neat thing about joining a Group is that you can monitor them and become a more active participant, post a comment yourself, or start a discussion. Currently you are limited to membership in fifty LinkedIn Groups. That number of groups should keep you busy, connected, and engaged!

- **Getting Personal References for Your LinkedIn Profile**

References are a tremendous way of demonstrating the value of your body of work to yourself, your customers, and your organization. A reference, in my opinion, is a reinforcement of your value proposition. And it is *earned*. I am not a big advocate of references for references' sake, as you will amass a large number of nondescript, uncontrolled references that look like "I have more references than you do." And all these references might end up saying is that you are a nice person. Put yourself in the shoes of your current or prospective employer or customer. What is the nature and context of the information you want to communicate to them about yourself? Ask for referrals and tell those people what you would like them to say about you if they are comfortable with that approach. You will be surprised by the quality and quantity of the referrals you receive.

TWITTER: www.twitter.com/yourname

Twitter® is a social commentary venue that is a good second tier strategy for technical/non-technical personal branding. Twitter allows users to post and read each other's updates, known as Tweets™. Tweets are limited to text posts of up to one hundred forty characters. As a result, Tweets are succinct. Like text messaging, there is a whole Twitter language that has developed to pack the most substance into one Tweet.

Tweets are posted on the author's profile for their entire network to see. Like LinkedIn, you can develop your own set of Followers based on your LinkedIn or Facebook connections, your connections strategy, and your personal brand. You can use the Twitter Search function to find people or identify groups by subject (e.g. aerospace engineers or online marketing experts). You can elect to "follow" people and they, in turn, can elect to follow you.

While some people build massive Twitter Followers lists and do not validate who elects to follow them, I recommend being prudent. Build out your network slowly. Scrutinize who is following you to make sure they are consistent with your brand and your values. Select relevant and consistent subject matter for your Tweets. And have fun! Introduce new topics related to your personal brand and include articles or blog posts you read, or even your favorite sports team.

A Twitter strategy takes a lot of time and a true plan. It is extremely sensitive to consistency and frequency of posting Tweets and the quality of those Tweets. Like all social networking strategies, you need to be authentic. Posting a Tweet once a week in conjunction with an event, update, or blog post may be all you have time for right now. Do your homework and have a strategy before you dive into this venue! Think before you Tweet.

FACEBOOK: *www.facebook.com*

Corporations and individuals with LinkedIn Profiles may also decide to list themselves on Facebook® as a means of building inbound links (a link from one huge, well-optimized site to another) to their LinkedIn Profiles. This strategy can help your company gain greater visibility on the Internet. Due to the highly variable nature of membership and the selective filtering of posted content, as well as recent security protocols, Facebook is not as porous for the search engines as LinkedIn, and that's probably a good thing. Facebook is a bit unwieldy to use as the fulcrum of your personal brand-building strategy. It helps if you already have a well-established follower base and, not surprisingly, a well-thought-out Facebook branding and business development plan.

Facebook, for most users, is intended for sharing the fun, personal side of who you are, even if you are a corporation with a Fan Page. If you are establishing a Facebook Fan Page, use your LinkedIn avatar, which can be your corporate logo, symbol, or personal picture, for consistency and co-branding. You can also include links to publications, blogs, and your website, and tap in to other creative means of growing your personal and professional brands. Corporations and niche affiliated marketers with well-thought-out personal branding strategies use Facebook as the main or an additional focus for market identification and business development.

Some corporations use Facebook for hiring if their target employment population uses Facebook more readily than job boards. If you know who your target population is and they're on Facebook, then you need to have a strategy to communicate with them. Hiring managers scour your online representation, so edit what you and your Friends are posting and what is visible online!

In February, 2011, Facebook introduced the "Secure Browsing" (HTTPS) option. As of October 1, 2011, Facebook mandated using a Secure Page URL to display to HTTPS users. Page tabs not hosted securely are no longer able to be displayed to users browsing under HTTPS. To learn more about how to create a Facebook Fan Page involving page tabs, start with "How to Move Your Facebook Tabs to Secure Hosting Required by Facebook," Tim Ware, http://www.socialmediaexaminer.com, September 30, 2011. (courtesy of Denise Wakeman, www.denisewakeman.com)

BLOGGING

A blog is a type of website maintained by an individual (the blogger) who ideally regularly posts commentaries about specific niche topics, descriptions of events, videos, etc., using various technology platforms. If you are building your personal or corporate brand (or both), and your plan includes establishing yourself as a thought leader, you should consider developing your own blog. Your blog should deal with a consistent and focused topic that reinforces your brand messages. Each blog post should be a reflection of your thoughts about specific topics relevant to your area of expertise.

If you are going to blog, do your own work and be ethical. Be original instead of a copycat. A formula that works for someone else isn't automatically a successful strategy for you. Tagging your blog post with another well-known entity's name in a post not related to that person, simply to drive traffic to your blog, only makes you look desperate. And it is a mandated blogging practice to give other bloggers credit, or attribution, when you cite their work.

People who blog are generally into mentorship; they want their posts to be syndicated and used by other folks. They want to catalyze and provoke discussion. Respect their work and the time and effort they put into blogging. Test out your blogging voice and create five draft posts before you publish. Have colleagues read them and offer their feedback. You may find out you are a natural at blogging, and that generating blog posts is an enjoyable, relaxing, and creative activity for you! You won't know unless you try.

REVIEW OF MAIN POINTS

1. Personal branding is an extension of your core values. Developing a personal brand allows you to communicate consistently, efficiently, and creatively with peers, employers, teachers, instructors, family members, and the thought leaders you engage with personally, professionally, and on social networking sites.

2. Developing a personal brand – at whatever age or career phase you are in – makes you a more valuable asset to yourself, your current or potential employer, and to your own enterprise, pre or post retirement. You may be starting a business venture and need investors, or perhaps you are changing careers.

3. Personal branding is about walking your talk. Personal branding involves, in the words of Dan Schawbel in *Me 2.0*: "authenticity, transparency, value, and visibility." Personal branding is the who, what, why, and where you are today in your career.

4. LinkedIn is a great place to start building your personal brand, followed by Twitter and then Facebook and/or Google+ if you already have a well-established network.

SALES-ENGINEERING INTERFACE™ TOOL #15: YOUR PERSONAL BRAND

Do I even have to tell you? If you don't have a LinkedIn Profile, develop one or add to/update your existing Profile. Is Twitter, Facebook, or Google+ your preference? Establish accounts. Connect and network. What are you waiting for?

CHAPTER SIXTEEN

TAKE CONTROL OF THE CUSTOMER ACQUISITION AND BUSINESS DEVELOPMENT PROCESS

Business Development Is a Process of Discipline and Alignment

No one is especially athletic when they first start a new exercise regimen, such as playing golf again at the start of each season, or learning Pilates. You may easily grow tired and rely on bad form – which results in poor outcome – just to complete the task. Eventually muscle memory kicks in and you are better able to execute, with better results. You gradually and successfully change your habits, and the positive results are tangible.

Business development involves the same sort of approach. You need to focus on the basics – the principles that make sense to you – as well as those that may not make sense at the start. You have to practice consistently in order to improve your performance. You may also have to invest in some equipment. In the case of learning about business development, fortunately a lot of training tools are free – or at least reasonably priced – and available online.

If you come from a technical venue, focus on incorporating solid business development practices naturally while implementing your engineering projects. If you are in sales, focus on gradually incorporating into your sales and marketing meetings your growing comfort with the technical aspects of your products and services; your objective is to consistently incorporate the technical aspects of the project as you seek to develop business, which eventually results in your closing the sale. Whether you are on the technical side of the table or the non-technical side, moving at least one millimeter outside your current comfort level is the overall goal. The most important requirement from you is your interest in and commitment to being more comfortable working both sides of your brain.

Keep in mind that you may never quite understand your client's motivations. No matter how well you feel you have aligned your project with your client's perspective, your client may still surprise you (or astound you) during the course of the project. Some clients will be negatively memorable. Others may have assured you that a project is going to materialize, but nothing develops after you spend a lot of time and effort with them. Then there are clients who make you reinvent the wheel through various iterations of the design and proposal, constantly changing its scope and telling you that they are talking to three other vendors as well as you. And yes, there are some prospective and current clients whom you will decide are not worth working with at all.

There are various scenarios that will present themselves to you during the course of your client-vendor relationships. Often when projects go wrong, you are able to look back in hindsight and realize the situation could have been averted or addressed early on in the project. It is your ability to see situations from a CEO mindset, using your new technical/non-technical collaborative business development perspective, that will allow you to collaborate with your clients towards long-lasting solutions.

Leverage What You Know against What You Don't Know

You are reading this book because you don't want to be regarded simply as an order-taker. You are interested in people asking your opinion rather than telling you what to do. As an innovator and leader, your thoughts have professional currency and value because you've earned the right to tender your opinions in the decision-making process. In that order-taker's role you were known to be risk-averse, and preferred to shift the responsibility to others.

By now you realize that no matter where you draw your personal line within your organization, business development remains a solid part of your job description. There is really no place to hide from your role in this process. It is better to acknowledge this reality and adapt, adopt, and apply the lessons we've addressed in this book into your professional responsibilities. You may be better at business development than you currently give yourself credit for.

If you want to differentiate yourself and have colleagues and clients asking you what you think, instead of whether you can do this or that task, then you have to develop the ability (and confidence) to weigh the factors impacting how decisions are made. Often during the course of a project, a team identifies a gap between what they know and what they don't know. And no one wants to admit that they don't have all the answers.

The entire project does not necessarily need to stop dead in its tracks once gaps are identified. Understanding what information is missing, and leveraging your ability to ask questions of the right people, is important in bringing a team together to solve the issues at hand. Utilize your LinkedIn Connections and thought leaders' articles to fill in the gaps. You may have more resources than you realize. Don't leave the asking of these questions up to someone else; they may be just as hesitant as you are to stick their neck out regarding information gaps. Kick this elephant out of the room so it won't keep you in your corner. Once the elephant has left the room, you may find that you are leading rather than following.

Being an innovator, rather than an order-taker or implementer, may involve nothing more than taking responsibility. In some corporate cultures this is a deviation from the norm. If you are not operating in the current status quo, you are being innovative. Being an innovator rather than an order-taker doesn't have to involve obtaining a patent for a new technical innovation; however, innovation can be perceived as the ability to bring people together to facilitate collaborative outcomes that speed up the order-to-cash process.

The ability to listen from the perspectives of both technical and non-technical disciplines, articulate these perspectives, and lead a collaborative discussion is a differentiator. You can't develop this ability if you are constantly fighting your colleagues, trying to push the sale of your product or service to customers, or wanting to rush off to design a solution based on what's been discussed thus far.

Innovation involves the wisdom of listening rather than speaking. It involves being able to translate the thoughts, perceptions, and mode of expression of one person to another person. It involves moving your mindset one millimeter outside your comfort level. Once you make that mental move, you will find that the way you see things changes. That is the type of responsibility you need to decide to take. It can make all the difference in the role you play in your organization and with your clients.

Do Not Expect Your Clients To Think in a Straight Line

People don't make decisions the same way you do; nor do they make them in a linear fashion. The traditional sales process chases prospects like a greyhound running after the racetrack rabbit. This process is based on the premise that if we ask questions in the proper order, the customer will answer them and come up with the obvious conclusion, which is to do business with us. By now I hope you understand that things rarely progress in this manner in the business development cycle. Don't get caught up in chasing that rabbit in a straight line to the finish when, in fact, you may end up taking a rather circuitous path in accomplishing your goals.

So who better to understand the non-linear business development process and decision-making algorithm than technically oriented sales, marketing, and engineering professionals?

What a thought! These individuals are constantly analyzing the multi-factorial and variable nature of the way things are. They are used to building into their designs and analyses responses to variances and machined tolerances to accommodate multi-factorial influences on outcomes, so they are already familiar with operating in a world of non-linear decision-making algorithms.

And if you are one of these technically-oriented professionals, have you ever considered that you may have a perceptual advantage in the business development process? You may be able to simultaneously see the big picture as well as the small details required in providing long-term, proactive, and robust solutions for your clients.

Understand Where Your Customers and Prospects Go

As you become a more confident, cross-functional business development professional, you will inevitably have to deal with the phenomenon of the disappearing customer. Sometimes your customers and prospects go away; you won't hear from them for weeks, or at all. Perhaps you were wearing blinders as you chased the rabbit around the racetrack towards what you assumed would be their straight-line decision to do business with you. Possibly you assumed that the sale or project was guaranteed.

Use your knowledge that people do not make decisions in a straight line to determine the reasons for this situation. Don't assume that you failed or that your customer is no longer interested. Just as your business changes on a day-to-day basis, so does theirs. Acquiring and retaining customers requires that you maintain contact with them even when you aren't doing business with them.

So where do your clients go when they go away? They promise to get back to you, but you never hear from them. You are embarrassed to contact them, but as time progresses, you begin to fear the worst. You feel you have failed. You blame yourself, your boss, and your company.

The one thing you don't want to do is call the client to find out where they have gone. You do have to contact them, though. Make sure you include relevant information that may be more pertinent to their business case than it is to your own agenda of winning their business. Use the following points to guide your conversation:

Are They Stuck?
The majority of the time, clients vanish either because they have become stuck in their own status quo or they are in a process of flux and change. Most of the time it's because your project or solution is perceived as being too disruptive to the status quo of their organization. They think they will have to change too many aspects of how they currently do business in order to bring your product or solution in house. They have elephants in their rooms, and there is no perceived place for your solution in these crowded spaces. What factors do you need to address to get them unstuck?

Are You Jumping through Their Hoops?
Sometimes your clients go away because they are more interested in making you jump through hoops than they were serious about the solutions and products you proposed. They are playing hard to get because you were anxious to please as you chased their business using straight-lined, traditional thinking. This strategy is demeaning to your company and to you. It brands your company as being vulnerable rather than savvy business developers. On the other hand, your prospect or client may treat all their vendors in this manner. Do you want to be associated with this type of company as your customer?

Are You Really Interested?
Sometimes your clients go away because you send them away. You push back from the table. You decide not to do business with a client or prospect because of a history of less-than-desirable outcomes. There may be an historically long sales cycle with lots of hoop-jumping. Perhaps you decide not to respond to yet another RFP because that is all you are to this company – a source for the third proposal. Having the confidence in yourself, your team, and your prospective or current customer to say, "Thanks, but no thanks" is an important asset to carry with you into any client meeting. It takes guts.

And while it's a drastic step, it may be important to prune back your customer base if it is full of these types of clients. In the long run, they create barriers to doing business with the types of companies you could do your best work for.

The business development process isn't a cosmic mystery. Stay focused and disciplined. Apply the questions you ask your prospects and customers to create a case for doing business with them. Their answers will gradually allow you to assemble what you feel their business model looks like; you can ask them more questions in order to make sure you see them the way they really are. And don't dismiss that your perceptions might provide valuable insights for them!

Consider where your solution or service fits in with how their company is organized. Determine whether this company is going to be a good fit for your products and services, or whether you are just using them as a means of achieving your sales or billing quota. Understand the timing of your business development meetings and whether they align with your customer's or prospect's timeline for finding a solution.

Doing this homework up front will save you a lot of wondering why your customer vanishes during the sales process. You will have the discipline to follow through. You will have the resources to fill in the gaps in your knowledge base about your customer. You will anticipate how they might react during various phases of the business development process. You may be able to avoid their going away because you are able to dispel their perception that your solution is too disruptive. And don't make assumptions. In spite of your excellent skill set and preparation, you *still* might inadvertently jump the gun and go for the close way too early in the business development process.

As you develop your ability to think like the customer, you will be better able to align and time your request for business. These are the same principles and methods that you apply to meetings with your internal colleagues. Developing the ability to simultaneously think on both sides of the table creates tremendous professional currency for yourself, your company, and, ultimately, your customers.

REVIEW OF MAIN POINTS

1. Developing a comfort level with the business development process is a matter of learning to think simultaneously like your customers and like the vendor that you are. While this process may move you one millimeter outside of your present comfort level, you are adapting, adopting, and applying to your external customer base the same principles you learned for communicating with your internal colleagues.

2. Are you an order-taker, an implementer, or an innovator? Do people ask, "Can you do this?" or do they ask, "What do you think about this?" Regardless of your present comfort level, business development *is* part of your job description. Becoming innovative may involve nothing more complicated than developing familiarity with the elements of a business plan and defining a business model.

3. No one makes decisions the same way you do, and they certainly don't make them in a straight line. There are multiple factors involved in determining what decisions your company and your customers need to make in order to place a solution or product. By determining these factors early on in the business development process and placing them within your perception of the customer's business model, you can shorten the overall order-to-cash process.

4. Your customers and prospects may go away during the business development process for a number of reasons. You need to determine these reasons rather than avoid thinking about them; the reasons may *not* be related to your failure as a business development professional. On the other hand, their decision to do business with another company, or not with anyone at all, may be the basis for improvement in your business development processes. Knowledge is constructive.

SALES-ENGINEERING INTERFACE™ TOOL #16: DO I REALLY KNOW MY CUSTOMERS?

1. Determine your role with your top five current customers. Are you an order-taker/implementer or an innovator?

2. Is the business development process lengthy with each of these customers? What is their style of doing business with you? Their style of doing business is a clue as to how they regard you and your company – order-takers or innovators.

3. Have these companies ever gone away during the course of the business development process? If so, where have they gone and for what reasons? If you don't know, find out.

4. What is your understanding of these go-away customers' business models? How can you determine their business models (online resources, internal resource at the customer, other reference material)? Are these companies a good fit for your own personal skill set? For your company's?

CHAPTER SEVENTEEN

PUTTING IT ALL TOGETHER

Business Development for Technical and Non-Technical Professionals

Regardless of whether you are a technical or non-technical professional, it's all about your ability to participate in all aspects of the business development cycle. Business development takes time and a lot of data – some qualitative, some quantitative, some technical, some non-technical.

It takes dialogue, and using those not-so-soft skills, to establish the context that generated the problem for which your client is seeking a solution. It takes understanding and appreciation of the people involved in the equation. Timing is all-important as well. And timing may not involve your ability to design and deliver the solution in a timely manner as much as it involves the time it takes for your customers and prospects to make that all-important decision to do business with you in the first place!

Business development involves understanding your current and potential markets and the customers within these markets, and developing products and deliverables to meet their needs. It is about how well you know your customers, their mindsets, and the context in which they make decisions. The sales cycle occurs about two-thirds of the way through the business development process. And it doesn't happen unless you've developed a business case for a solution.

Where did you fit into this business development equation before you read this book? And where do you fit into these dynamics now that you are learning to sit on both sides of the table and simultaneously translate ideas and insights?

No one comes out of engineering school or business school with the full package. Somewhere along the line individuals decide to stop waiting for someone to throw them a lifeline. They start to fine-tune their expertise, just as you are doing by reading this book. They cross-train their brains, just as you are doing by engaging in collaboration along the sales/engineering interface. And they begin to have confidence that they can self-direct their careers.

Are Technical Professionals the Stewards of the New Business Development Paradigm?

Take the time to familiarize yourself with the elements of a business plan. If you are a technical professional, you may find that this exercise provides you with a far greater comfort level and a lot of data from which to communicate with your customers. If you can talk business data, you are speaking the language of the business owner. If you can discuss input, output, and throughput, and relate it to profit and loss, you have something significant to communicate to a decision-maker. If you can relate this information to industry data and quantitative market trends, I'd say you have the basis for creating a compelling reason for others to do business with you.

If you are a sales professional, do yourself a favor and develop a stronger interest in the data-driven aspects of technology and industry trends. Having the ability to discuss business with your prospects and customers from their perspectives builds credibility. Letting go of your sales-training habits, which tell you to constantly identify pain points so that you can shift the discussion to a selling mode, may be the most beneficial transition you can make in moving outside of your comfort level. Become a business development professional. Why confine yourself to the sales process?

With so much information available, business development is calling for those companies and individuals who know how and where to gather information, collaboratively interpret this information, and translate it for colleagues and clients. Business development includes specifying those individuals who are willing to think with both sides of their brains and simultaneously sit on both sides of the table.

Your ability to make business development and your core values a part of your daily habits can fuel your personal and professional development. Synthesizing rather than compartmentalizing the constant stream of information that comes your way in this digital age makes you a more valuable resource and colleague.

Bring Your Personal Core Values into Every Professional Interchange

Throughout everything, you bring your personal core values – the ethics and value system that form the foundation of who you are – into all you touch. And your core values center you; you can always return to them when things get messy and stressful. Your cross-functional skill set and professional core capabilities are grounded in your personal core values.

We all have habits – our own status quo. Some of them are good, and some bad. Perhaps the worst habits we have involve self-doubt and second-guessing ourselves. Until we stop the voices inside our heads that get in the way of where we want to go, we will stay firmly stuck in our own status quo – our own version of the elephant in the room. Realign yourself with your personal core values. Focus on your core values to fuel your passion, talent, and gifts.

Your personal core values are not some form of career- or money-driven professional goal, nor are they your job description, or even your function. They are the path you take in life; the people you meet along the way who become a significant part of your life, your career, and your personal choices; and the things you do on behalf of others. Jim Collins, in *Good to Great: Why Some Companies Make the Leap... and Others Don't*, characterizes the success of major companies as revolving around the efforts and vision of individuals who possess passion, talent, and gifts. I would respectfully like to add a fourth trait to this mix: well-developed personal core values.

It Won't Be a Piece of Cake, But If It Were, It Wouldn't Be Fun

I've planted a lot of ideas, and given you tools with which to organize your thoughts and a structure you can constantly revisit to implement what you've learned. The toughest decision I am now asking you to make is to move one millimeter outside of your current comfort level.

It is hard to change habits, switch gears, and unseat our biases. These changes don't happen overnight. However, the marketplace and global economy are exerting daily pressures on our professions and the companies for which we work. Change is inevitable. Your ability to proactively retool and recalibrate your core capabilities in anticipation of these changes will permit you to be better prepared for what's down the road.

What I am asking you to do – to incorporate what you learned in this book and from the Sales-Engineering Interface™ toolkit and tips – is not easy. You will succeed in some situations and fail in others. You will doubt yourself and your abilities. You will want to crawl back to your cubicle and hide. If you are a salesperson, you will run back to the phone, dialing for dollars and churning and burning your way through leads lists of supposedly qualified prospects. You will even try to revert back to your former status-quo habits, mindset, and biases! But they won't feel comfortable anymore…because change is inevitable, and you have shifted your perspective ever so slightly – one millimeter outside of your comfort level, to be exact.

The status quo is a comfortable place to hide. But it is not a comfortable place for growing. Leave self-criticism and taking things personally behind. Moving forward involves trying new things. Like learning to ride a bicycle, you are going to fall off and get bruised more than a few times. You can't just try things once, fail, and dismiss them as not having value. It's like reading a book about how to run a marathon and then going out the next day expecting to run a full race. You need to be realistic and pace yourself. You will find that you become intrigued and energized by changing things up and moving forward. After all, you are seeing the situation in a new perspective and expressing yourself differently, and people are going to be reacting to you differently. And more productively.

You are responsible for re-engineering your career. Understand what works easiest for you and why. Which of the tools that I've given you are the hardest to incorporate into your habits? How can you gradually refine your approach to ease these concepts into your workday scenarios? At first you will feel like you need to constantly run back to this book and these tools before you make a move or have a discussion. Gradually you will find yourself communicating, introducing ideas, and asking questions naturally within the new structure of your business development mindset. People may ask you why you are smiling as you are speaking with them. You'll know the answer, and simply continue to smile. You have moved one millimeter outside of your comfort level!

You are always a work in progress, and your goal isn't to be perfect. Your colleagues, prospects, and customers – and their businesses – are in the same boat. They are asking the same questions you are, and experiencing the same successes and frustrations that you do. Your goal is to be so tuned-in to the dynamics of the business development process that you become their resource for insight from this technical/non-technical perspective. Business development occurs within a constantly shifting environment. There is no stasis, ever. It requires that individuals constantly collect the data and have those proactive and collaborative discussions that keep them one or two product life cycles ahead of the competition.

Start incorporating this dynamic mindset into low-risk conversations and parts of small projects. Gradually build your repertoire and comfort level. Don't go for the jugular from the start. Subscribe to RSS feeds about industry trends. Learn who are the voices in your industry and with whom you might want to work. Who are the movers and shakers, and why are they perceived in this manner? You will be surprised to find that their ideas are harmonious with the goal you are moving towards. As you incorporate your expanded knowledge base into your business development conversations, you will start to emerge as the cross-functional go-to guy or gal. Take the risk to move that one millimeter outside of your comfort level. That's all it takes. The rest will follow. And you will find that *you*, definitely, do mean business.

STAYING FOCUSED

1. You've already taken the step to move one millimeter outside of your comfort level by reading this book. Remember, you are a work in progress. Continue!

2. Don't go for perfect – go for changing things up. If you feel uncomfortable, you are colliding with an ingrained habit. You don't have to throw everything out, just the stuff that holds you back.

3. Practice, practice, practice! The business environment is dynamic. You won't be saying the same things to people all the time because contexts change. Be aware of this factor and use it to ground your business development conversations.

4. Communicate with colleagues to determine how their contexts and needs have changed. Prepare for meetings with customers' (internal and external) needs in mind. Use the industry and technology (or other) information you are reading to provide greater depth and insight to meetings, conversations, and reports.

5. Asking for help isn't a sign of weakness or lack of knowledge. It's a sign of self-belief, confidence, and a willingness to improve. You have great resources in your internal colleagues, mentors, and professional online communities. Tap into the thought leaders – they are tremendously accessible.

6. Leadership is about willingness to assume responsibility and accountability. If you are one millimeter outside of your comfort level, you've already moved well beyond the cubicle mindset. That shift in itself has you leading instead of following. Carry on!

7. Adapt, adopt, and apply your business development mindset and cross-functional perspective until they become second nature to you, personally and professionally.

8. No matter whether you are a technical or non-technical professional, collaborative business development *is* part of your job description.

What are you waiting for?

ABOUT THE AUTHOR

BABETTE N. TEN HAKEN helps technically oriented manufacturing, IT, and engineering service companies and entrepreneurs generate revenue. She has a firm belief that business development is fueled by a non-siloed business model that creates synergy between technical and non-technical professionals. Today's successful owners, employees, and entrepreneurs need to be able to collaborate across disciplines and confidently incorporate upstream and downstream business information into their job functions. It's about providing value – to yourself, your clients, and your company.

Her popular **Sales Aerobics for Engineers® blog** is a destination for technical and non-technical professionals, recent graduates, business owners, and entrepreneurs who are stuck in the status quo and want to move to the next level of client acquisition and revenue generation. Find it at http://salesaerobicsforengineersblog.com.

Babette's **Sales Aerobics for Engineers®** consulting, training, and coaching efforts engage technology- and engineering-intensive manufacturers and service companies that are looking for ways to expand their deliverables to current customers and penetrate new markets. She is called on to facilitate meetings between R&D and marketing/sales professionals, and just about any other us-versus-them situation, providing the *simultaneous translation* skill set that results in all parties seated around the table speaking productively and collaboratively. Her ability to correlate processes and practices with an understanding of the dynamics of revenue generation allows her clients to have a more focused, consistent, and disciplined approach to business development. Her company website is www.salesaerobicsforengineers.com.

Babette Ten Haken holds a dual degree in physical anthropology and evolutionary genetics from Washington University in St. Louis, Missouri, USA. She earned her Masters Degree from Heythrop College, University of London, UK, receiving First Honors for her work in the history of science and the interface between theology and science. She is a Six Sigma Green Belt and a Certified Voice of the Customer Coach. She is passionate about business development and cross-functional collaboration.

RESOURCES

BLOGS

Babette Ten Haken, author, consultant, and blogger, http://salesaerobicsforengineersblog.com. Focuses on the value of business development based on the interface between sales and engineering, and technical/non-technical collaboration.

Jill Konrath, strategist and author, www.jillkonrath.com (formerly www.sellingtobigcompanies.com). Expert account entry strategies for sales professionals.

Katie Konrath, blogger, www.getfreshminds.com. As Katie says, "Ideas so fresh...they should be slapped!"

Civil Engineering Central, http://blog.civilengineeringcentral.com, the blog of the Civil Engineering Central LinkedIn Group and Civil Engineering Central recruiting, specializing in architecture and civil engineering, and authored by a team of regular guest bloggers including Matt Barcus, Carol Metzner, Babette Ten Haken, and Anthony Fasano, among others.

Seth Godin, author, consultant, and blogger, *Seth's Blog*, http://sethgodin.typepad.com. The guru of marketing shares his wisdom, expertise, and insights on human nature and everything else.

Personal Branding Blog, http://personalbrandingblog.com. Dan Schawbel's blog forum, authored by Dan and a team of regular guest bloggers, providing insightful information about *Me 2.0* and personal branding.

Anthony J. Fasano, consultant and author, www.powerfulpurpose.com/blog, focusing on career development for engineering professionals.

Patrick O'Malley, author, blogger, and consultant, *Social Media Super Blog*, http://www.the-linkedin-speaker.com/blog, focusing on how to fine-tune your social media content to make it work for you more effectively.

Mike Shipulski, PhD, professional engineer and blogger, *Shipulski on Design*, www.shipulski.com. Product development, product design, and engineering.

Keith Sawyer, PhD, professor, author, and blogger at *Creativity and Innovation*, http://keithsawyer.wordpress.com. A scientist who studies creativity and tells you where creativity happens in business, culture, and technology.

Sharon Drew Morgen, author, coach, and blogger, http://sharondrewmorgen.com, facilitating the entire buying-decision path.

Anthony Parinello, author, coach, and blogger at *VITO™ Selling, Tony's Blog*, www.vitoselling.com/category/blog. Online site based on his best-selling book about how to sell to the top officer in the organization.

Sam Richter, author, consultant, and blogger, www.knowmoreblog.com. Focusing on effective selling techniques incorporating information gathering about clients, prospects, industry trends, and trigger events.

SOURCES OF QUOTES AND BOOKS CITED, BY CHAPTER

Chapter 1

Keith Sawyer, *Group Genius: The Creative Power of Collaboration*, Basic Books, 2007

Harvard Business Review, *Making Smart Decisions*, Harvard Business Review Press, 2011. "The Hidden Traps in Decision Making," John S. Hammond, Ralph l. Keeney, and Howard Raiffa; "Conquering a Culture of Indecision," Ram Charan; "Evidence-Based Management," Jeffrey Pfeffer and Robert I. Sutton; "How (Un)ethical Are You?" Mahzarin R. Banaji, Max H. Bazerman, and Dolly Chugh

Chapter 3

USA Today, "Jacks of All Trades, and Masters of All," by Paul Davidson, Wednesday, July 6, 2011

Keith Sawyer, *Group Genius: The Creative Power of Collaboration*, Basic Books, 2007, p 131

Chapter 4

Jill Konrath, *Selling To BIG Companies*, Kaplan Publishing, 2006

Jill Konrath, *SNAP Selling: Speed Up Sales and Win More Business with Today's Frazzled Customers*, the Penguin Group, 2010

Chapter 6

National Engineering Week Press Release, 2011, "The Root of Ingenuity – The Engineer," National Engineers Week Foundation Press Release, http://www.eweek.org/site/News/Features/root.shtml

Seth Godin, Seth's Blog, "Merging/Emerging," http://sethgodin.typepad.com/seths_blog/2011/09/mergingemerging.html

Chapter 7

Dilbert® comic strip, Scott Adams, Cartoonist

"Seventy-One Percent of Employers Say They Value Emotional Intelligence over IQ, According to CareerBuilder Survey," http://www.careerbuilder.com, August 18, 2011

Chapter 8

Jim Collins, *Good to Great: Why Some Companies Make the Leap... and Others Don't*, HarperCollins Publishers, 2001

Thomas L. Friedman, *The World is Flat: A Brief History of the Twenty-first Century*, Farrar, Straus, and Giroux, 2007

Chapter 9

Tom Peters, "The Brand Called You," *FastCompany.com*, August 31, 1997, http://www.fastcompany.com/magazine/10/brandyou.html

Dan Schawbel, *Me 2.0: Build A Powerful Brand To Achieve Career Success*, Kaplan Publishing, 2009

Anthony Parinello, *Selling to VITO™, the Very Important Top Officer*, Adams Media Corporation, 1994

Chapter 10

SCORE, at www.SCORE.org. A resource for templates and assistance for small businesses.

U.S. Small Business Administration, www.sba.gov. Another resource for templates and assistance in forming a small business.

Chapter 11

Sam Richter, *Take the Cold Out of Cold Calling: Web Search Secrets*, 3rd Edition, Beaver's Pond Press, 2009

Chapter 14

Jill Konrath, *Selling To BIG Companies*, Kaplan Publishing, 2006

Jill Konrath, *SNAP Selling: Speed Up Sales and Win More Business with Today's Frazzled Customers*, the Penguin Group, 2010

Sam Richter, *Take the Cold Out of Cold Calling: Web Search Secrets*, 3rd Edition, Beaver's Pond Press, 2009

Chapter 15

Dan Schawbel, *Me 2.0: Build A Powerful Brand To Achieve Career Success*, Kaplan Publishing, 2009

"Seventy-One Percent of Employers Say They Value Emotional Intelligence over IQ, According to CareerBuilder Survey," http://www.careerbuilder.com Press Release, August 18, 2011

"How to Move Your Facebook Tabs to Secure Hosting Required by Facebook," Tim Ware, http://www.socialmediaexaminer.com, September 30, 2011

Chapter 17
Jim Collins, *Good to Great: Why Some Companies Make the Leap... and Others Don't*, HarperCollins Publishers, 2001

GLOSSARY OF TERMS, BY CHAPTER

Unless otherwise noted, sources include:
www.onlineslangdictionary.com
www.businessdictionary.com
www.freedictionary.com
www.webster-dictionary.org
www.merriam-webster.com

Chapter 1

Techie – slang, a person who is very knowledgeable or enthusiastic about technical and hi-tech subjects.

Geek – slang, an enthusiast or expert, especially in a technological field.

Nerd – an unstylish or socially inept person, especially one slavishly devoted to intellectual or academic pursuits.

Group synergy – a mutually advantageous collection or compatibility of business participants or elements.

Business development – a specialist area of business involving a number of techniques and responsibilities that aim at penetrating existing markets and attracting new customers.

Job description – the general tasks, functions, and responsibilities of a position for which an individual is hired.

Job functionality – how the routines, processes, and skills of an individual relate to the job they were hired to do.

Throughput – work from one department or discipline that is handed off to, or put through to, another department.

Output – what is produced.

Root cause – a single cause for an outcome that, if prevented, would prevent the outcome itself. In this context, the root cause is the cause that dominates over all other contributing factors. (www.asq.org)

Departmental silos – a business model, the name of which is a metaphor suggesting a similarity between grain silos that segregate one type of grain from another, and the segregated parts of an organization. In an organization suffering from silo syndrome, each department or function interacts primarily within that silo rather than with other groups across the organization. ("Smashing Silos," Evan Rosen, *Bloomberg Businessweek Blog*, February 5, 2010)

Cross-functional – work across disciplines.

Collaboration – working with others, especially in an intellectual endeavor.

Chapter 2

Naysayer – one who denies, refuses, opposes, or is skeptical about something.

Decision-making – the mental process of considering and reaching a decision, position, opinion, or judgment.

Risk - the potential that a chosen action or activity (including the choice of inaction) will lead to an undesirable outcome.

Interdepartmental – involving or representing more than one department of a business, academic institution, or government.

Engineer – an individual who is trained or professionally engages in a branch of engineering, or who operates an engine, or who carries through an enterprise using skillful or artful contrivance.

R&D (research and development) – the part of an enterprise's activities concerned with applying the results of scientific research to develop new products or services and improve existing ones.

Bean counter – slang, an accountant, or a bureaucrat who is believed to place undue emphasis on the control of expenditures.

Marketing – the act or process of buying and selling in a market, or the commercial functions involved in transferring goods from producer to consumer.

Sales – the exchange of goods or services for an amount of money or its equivalent; the act of selling.

Operations – the division of an organization that carries out the major planning and operating functions.

Cash flow – the movement of money into and out of a business.

Profitability – the quality of affording gain, benefit, or profit.

Billing cycle – the period of time between billings or invoicing.

Customer – one who regularly or repeatedly makes purchases from a vendor.

Interface – a shared boundary, or the point at which independent systems or diverse groups interact, for instance an academic discipline interacting with a corresponding department in a business.

Scenario – a postulated sequence of events.

Infrastructure – the basic structure or features of a system or organization.

Sales engineer – an individual who provides technical support to individuals selling the products or services of an enterprise.

Sales quota – the minimum sales goal that a sales agent is expected to achieve for a specific period of time.

Rework – make over.

Value-add – a perceived increase or enhancement of the quality of a product or service provided by a vendor that results in customer satisfaction, repeat or renewed business, or the decision to specify that vendor.

Vendor – an enterprise or person who regularly or repeatedly sells products or services to customers.

Chapter 3

Common denominator – a common trait or theme.

C-level decision-makers – company officials and executives whose title usually begin with the letter *C*, such as CEO (chief executive officer) or CFO (chief financial officer).

Business-to-business (B2B) – businesses whose customers are also businesses.

Business-to-consumer (B2C) – businesses whose customers are consumer end users rather than other businesses.

Revenue stream – uninterrupted receipts of revenues.

Quantitative data – data measured or identified on a numerical scale that can be analyzed using statistical methods.

Tolerances – an engineering and mechanical engineering term that means the permitted variation in some measurement or other characteristic of an object or work piece.

Marketing research – the function that links the customer and the public to the marketer through information used to identify and define marketing opportunities and problems.

Quantitative survey methodology – the sampling of individuals from a specific or general population with a view towards making statistical inferences about the population using the sample. (www.socialmarketing.com)

DMAIC – an initialism that represents a five-step approach for driving costly variation from manufacturing and business processes. The initials stand for define, measure, analyze, improve, and control. The backbone of the Six Sigma methodology, DMAIC delivers sustained, defect-free performance and highly competitive quality costs over the long run. (http://www.dmaictools.com)

Hypothetico-deductive model or scientific method – a method of investigation involving observation and theory to test scientific hypotheses introduced by William Whewell in *History of the Inductive Sciences*, 1937

Extrapolation – inference about the future based on known facts and observations.

Left brain – the left hemisphere of the human brain, controlling activities on the right side of the body, and which is believed to control linear and analytical thinking, decision-making, and language.

Right brain – the right hemisphere of the human brain, controlling activities on the left side of the body, and usually controlling perception of spatial and nonverbal concepts. The thought processes involved in creativity and imagination are generally associated with the right brain.

Chapter 4

Brand – a trademark or distinctive name identifying a product, service, enterprise, or individual; a product line identified by a trademark; a distinctive category or individual.

Techno-speak – a term used by Babette Ten Haken to signify technobabble (a combination of the words technology and babble), which is a form of speech using jargon, buzzwords, esoteric language, specialized technical terms, and/or technical slang that is incomprehensible to the listener. (Wikipedia)

Business babble – a term used by Babette Ten Haken to describe a form of speech using jargon, buzzwords, esoteric language, specialized business terms, and/or business slang that is incomprehensible to the listener.

Point load distribution – a means of calculating the load-carrying and deflection characteristics of beams. (Timoshenko, S., (1953) *History of Strength of Materials*. McGraw-Hill, New York)

Truss system – an architecture and structural engineering term referring to a rigid framework of wooden beams or metal bars designed to support a structure such as a roof or bridge.

Lingo – the specialized vocabulary of a particular field or discipline.

Semantics – the meaning or the interpretation of a word, sentence, or other language form.

Value proposition – a clear statement of the tangible results you expect a customer to get from using your products or services. It is focused on outcomes, and stresses the business value of your offering. (Jill Konrath, *Selling to BIG Companies*, p. 51)

Elevator speech – a short one- or two-sentence statement defining who you work with and your value proposition (see above). (Jill Konrath, *Selling to BIG Companies*, p. 52)

Business spiel – another way of referring to business babble and elevator speeches; a speech that uses a lot of business-related terms as a means of promoting the sales process.

The real deal – slang, something legitimate.

Chapter 5

On-boarding – employee orientation or organizational socialization of newly hired employees. (www.careerbuilder.com)

Core capabilities or core competencies – a specific factor or factors that a business or person sees as being central to the way they work. Core capabilities are not easily imitated by competitors or colleagues, can be leveraged widely to many products, markets, and output, and contribute to the end user's experienced benefits.

Commoditize – to perceive a product, service, discipline, or person as indistinguishable from, or the same as similar types of products, services, disciplines, or people.

Stereotype – lacking individuality or originality.

Chapter 6

Engineer – "John Lienhard, professor of mechanical engineering and history at the University of Houston and host of National Public Radio's *Engines of Our Ingenuity*, traces the word engineer to the Latin word *ingeniare*, which means to devise. Several other words are related to this word, including ingenuity." (National Engineering Week Press Release, 2011, http://www.eweek.org/site/News/Features/root.shtml)

Professional currency – the perceived worth of your value proposition and your core capabilities in the eyes of your internal and external customers.

Customer retention – the activities that a selling organization undertakes in order to reduce customer defections.

New business acquisition – the strategy, process, and activities involved in winning business from customers with whom your company has not done business in the past.

Discrete problem-solving – the techniques, methodologies, and formulas use to solve specific types of problems related to specific types of professional disciplines.

Billable hours – employee work that can be directly attributed to and invoiced to specific customer projects.

Non-billable hours – employee work that cannot be attributed to or invoiced to specific customer projects.

Multi-disciplinary – crossing many professional disciplines.

Chapter 7

Soft skills – communication strategies involving listening, speaking, presenting, writing, or knowledge of the context and history of a business, including events, people, successes, failures, and management, and leadership skills such as accounting, planning, strategy, and ethics.

Tweet – a short message sent via www.twitter.com, consisting of 140 characters or less.

ACT – a standardized achievement examination for college admissions in the United States produced by ACT, Inc.

SAT – Scholastic Aptitude Test – a reasoning test standardized for college admissions in the United States.

GMAT – Graduate Management Admission Test – a computer-adaptive standardized test in mathematics and the English language used for measuring aptitude to succeed academically in graduate business studies.

Chapter 8

Employment continuum – the progression of jobs within an enterprise or corporate environment that an employee can achieve.

Newbie – slang, an individual newly hired into a business or technical position.

Best in class – The highest current performance level in an industry, used as a standard or benchmark to be equaled or exceeded. Also called "best of breed."

MBA – Master of Business Administration, a graduate business degree.

Contracted worker – a self-employed individual or independent contractor who performs a specific job for a targeted period of time.

Entry level position – bottom level employment usually requiring minimal education, training, and experience.

Chapter 9

Internal customer – an employee who receives goods, services, or output produced elsewhere in their organization.

External customer – a party or enterprise that uses or is directly affected by a company's products or services.

Core values – principles that guide an organization's or individual's internal conduct and relationship with the external world. An organization's core values are usually summarized in its mission statement or in a statement of core values.

VITO™ – an acronym for "Very Important Top Officer" created by Anthony Parinello in his book *Selling to VITO™, the Very Important Top Officer*.

The big picture – slang, everything involved in a situation, especially the long-term or unforeseen repercussions of a situation that is often more narrowly defined.

Chapters 10, 11, 12

Business plan – a formal statement of a set of business goals, the reasons why they are believed attainable, and the plan for reaching those goals. It may also contain background information about the organization or team attempting to reach those goals.

Strategic plan – an organization's process of defining its strategy or direction and how it will allocate capital and human resources to pursue the strategy.

SCORE – a nonprofit association dedicated to educating entrepreneurs and helping small businesses start, grow, and succeed. (www.SCORE.org)

Small Business Administration – the official government-sponsored entity set up to support small businesses across the United States of America (www.sba.gov).

Executive summary – sometimes known as a management summary – a short document or section of a document, produced for business purposes, that summarizes a longer report.

Business model – a description of how a company organizes its capital and human resources to deliver output to the end user and accrue revenue, including the rationale for the process.

Revenue model – the way a business makes money from the sale of its products and services.

Chapter 13

Sales process – also known as a sales tunnel or a sales funnel – a systematic approach to selling a product or service.

A-list – a list or group of individuals at the highest level of excellence or eminence in an organization or society.

RFP – request for proposal

Chapter 14

Sales funnel – also known as a sales tunnel or a sales process – a systematic approach to selling a product or service.

Talking head – a person who is empty and pretentious. Originally used as a reference to a newscaster on television. (www.dictionary.reference.com)

Cold calling – the practice of contacting prospective customers who were not expecting such an interaction, in order to create business.

Chapter 15

Personal branding – the process by which individuals and entrepreneurs differentiate themselves by identifying and articulating their unique value proposition, whether professional or personal, and leveraging that value proposition across platforms with a consistent message or image to achieve a specific goal. (Dan Schawbel, *Me 2.0: Build A Powerful Brand To Achieve Career Success*)

Digital millennium – the era created by The Digital Millennium Copyright Act (DMCA), a United States copyright law that implements two

1996 treaties of the World Intellectual Property Organization (WIPO). The law criminalized production and dissemination of technology, devices, or services intended to circumvent measures (commonly known as digital rights management (DRM)) that control access to copyrighted works. It also criminalized the act of circumventing an access control, whether or not there is actual infringement of copyright itself, and heightens the penalties for copyright infringement on the Internet. (www.copyright.gov/legislation/dmca.pdf)

Chapter 16

Innovator – a person who begins or introduces something new.

Order-taker – an individual who follows orders.

Cosmic mystery – slang, a situation that is so broad in scope or complex as to have no logical means of explanation.

Chapter 17

Avatar – a picture or photo, 128 x 128 pixels in size, that is used on the Internet to identify a specific person (Dan Schawbel, *Me 2.0: Build A Powerful Brand To Achieve Career Success*, p. 214)

CPSIA information can be obtained at www.ICGtesting.com
Printed in the USA
LVOW011944161212

311893LV00016B/260/P